IPHONE 13 USER GUIDE

The Easy & Complete Manual to Discover All iPhone 13 Tips & Tricks For Seniors & Beginners. Discover All You Can Do with Your Smartphone, Apart from Calling, & How to Do It!

Mike Wang

Mike Wang

© Copyright 2021 - All rights reserved.

This document is geared towards providing exact and reliable information in regard to the topic and issue covered.

- From a Declaration of Principles which was accepted and approved equally by a Committee of the American Bar Association and a Committee of Publishers and Associations.

In no way is it legal to reproduce, duplicate, or transmit any part of this document in either electronic means or in printed format. All rights reserved.

The information provided herein is stated to be truthful and consistent, in that any liability, in terms of inattention or otherwise, by any usage or abuse of any policies, processes, or directions contained within is the solitary and utter responsibility of the recipient reader. Under no circumstances will any legal responsibility or blame be held against the publisher for any reparation, damages, or monetary loss due to the information herein, either directly or indirectly.

Respective authors own all copyrights not held by the publisher.

The information herein is offered for informational purposes solely and is universal as so. The presentation of the information is without contract or any type of guarantee assurance.

aAll trademarks and brands within this book are for clarifying purposes only and are owned by the owners themselves, not affiliated with this document.

TABLE OF CONTENTS

Introduction   9
What is the difference between the iPhone 13   13
   Features of iPhone 13   13
   Tips and tricks of iPhone 13   24

Features of iOS 15   33
   Facetime   33
   Messages   33
   Memoji   34
   Focus   34
   Notification   34
   Map   34
   Safari   35
   Live text   35
   Visual lookup   35
   Memories   35
   Reminders   35
   Weather   36
   iCloud+   36

Set up iPhone 13   37
   Understanding your options   37
   Choose your preferred language   38
   How to wake and unlock iPhone 13   38
   How to unlock iPhone with Face ID   39
   How to Turn Off Notification on your iPhone 13   40
   Change Notification Settings from an Incoming Notification   41
   Turning Off All Notifications on your iPhone 13   42
   What Happens When you Switch Off Notification on Your iPhone   42
   How to Pair Hearing Aids to iPhone 13   42
   Back up iPhone with iCloud   43
   Explore 5G   47
   Data and Voice   48
   Data Mode   48
   Link iPhone to Wi-Fi network   49
   Connect to a Personal Hotspot   50
   Connect your iPhone to a mobile network   50
   How to Setup your iPhone 13   50

How to set up Apple Pay — 51

How to confirm identity for Apple Pay — 51
Set Up Apple Pay — 51
How to Make Apple Pay More Convenient When Shopping in Apps — 54
Modify the Default Shipment and Contact Information for Your Account — 55
Use Apple Cash — 55
Use Apple Card — 57
Pay Your Bills — 59

Explore the App Store — 61

Improve the reading conditions at Night — 64
Restrict the access to certain Books — 64
iTunes — 65
Create a Focus — 67
Customize Focus — 68
Enable Time-Based Notifications — 69
Activate Smart Activation — 69
The Settings App — 70
Notifications — 71
Features under "Display and Brightness" settings — 73

iPhone 13 — 77

How to Force Close Active Applications on the iPhone 13 from the App Switcher — 77
How to create a folder on the Home Screen — 78
Widgets on the Home Screen — 79
Wake and unlock your iPhone — 80
Restart or Turn off the iPhone — 80

Apple ID and iCloud — 83

Transfer Data from an Android Device to your new iPhone — 84
Everything about Widgets on Home Screen — 85
Building a Smart Stack — 86
How to Take a Screenshot — 87

Everything about FaceTime Call — 89

Using Facetime on the iPhone, Take a Live Photo — 89
Use Other Applications During a Facetime Conversation — 90
Begin a Facetime Call with a Group of People — 90
Creates a Blurred Background in Portrait Mode — 91
How to Drag and Drop screenshot — 92
How to Focus — 92
How to Get Augmented Reality Walking Directions in Maps — 95
How to Use Siri to Share Whatever Is on Your Screen — 96
How to Use Apple Maps to Locate Transit Stations Near You — 97

Use your iPhone or iPad to launch FaceTime. 98
Unlocking with an Apple Watch Isn't Working? Here's How to
Solve the Issue 99

Siri 101

How to Instruct Siri to Control Your HomeKit Devices at a
Predetermined Time 101
How to Make Siri Read Your Notifications 102
How to Make a Private 'Hide My Email' Address 102
Make the Most of Apple's New Weather Maps 103
How to Use the Translate App's Auto-Translation 104
How to Refresh a Webpage Quickly in Safari 104
How to Change the Start Page and Background of Safari 105
How to Make Use of Tab Groups in Safari 106
Inquire of Siri 106
Utilize Your Voice to Activate Siri 106
Add Shortcuts for Siri 108
Charge iPhone With MagSafe Charger 109
Connect Your Apple Watch to Your iPhone 110

Handoff Work Across Your iPhone and Other Devices 111

Prior to Beginning 111
Handoff from a Different Device to Your iPhone 112
Handoff From The iPhone to Another Device 112
Disable Handoff on All of Your Devices 113
Between Your iPhone and Other Devices, Cut, Copy, and Paste 113
Copies, Cuts, and Pastes 114
Sync Your iPhone to Your PC 114
Calendars and contacts 115
How to Send Content from an iPhone to a Computer 116
Connect your iPhone and a computer using the USB cable
for a variety of reasons. 117
Handoff 117
Erase iPhone 118
Restore to iPhone to Default Settings 119
Back up your iPhone 120
Apps and Features of the iPhone 120
Airdrop 121

How to Use Hide My Email to Create an Email
Address 123

How to Keep Your IP Address Private While Using Safari 124
How to Activate and Deactivate iCloud's Private Relay 126

Conclusion 129

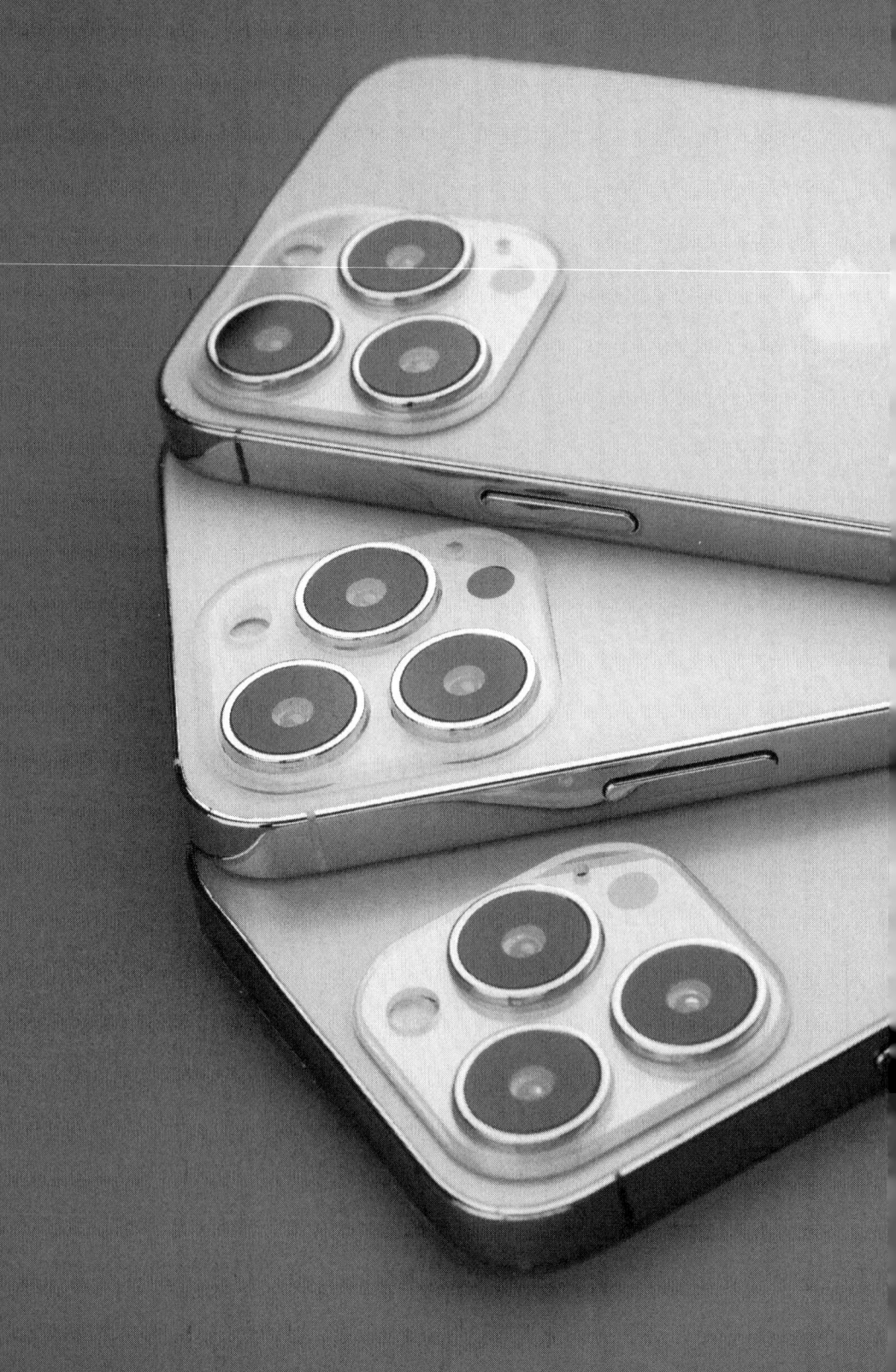

INTRODUCTION

Each year, like clockwork, we see a new iteration of Apple's extremely popular smartphone. This period, it's the time of the iPhone 13 series devices, and the organization is making discreet alterations to the previous year's products.

iPhone will be similar to the composition of the 2020 iPhone, with four devices including 6.1 inches, 6.7 inches, and 5.4 inches, two iPhones with greater-end Pro models, and two placed as low-priced, with very inexpensive gadgets.

Significant design modifications are not much anticipated, but the 2021 iPhones will come with similar functionality as the way 2020 iPhones are, and with a camera upgrade, and with a quicker A-Series processor, Qualcomm's novel 5G chip, battery life enhancements but it fixes, and all have been sized from the scratch.

In light of that, there are just little contrasts between the 2021 models and that of the 2020 models. However, there are still four registered iPhones, which can be confusing for the average shopper.

Besides some unobtrusive contrasts, you will not see many alterations in the current iPhone devices similar to what happened last year, there are four models, and each model is generally a similar size as the earlier year's same. The number of cameras on the back of each device remains the same, although they have been slightly redesigned on entry-level models. The face of the gadgets is the thing that has appeared differently the most, with the well-known indent that was refined interestingly.

The vanilla version is the usual and average model. It has the same screen size as last year, 6.1 inches. It likewise has a similar back camera arrangement, with a principal shooter matched with a super wide-point focal lens.

The normal iPhone 13 and little version, which is the mini model Apple's freshest and most reasonable iPhones, and is sold close by the more costly Pro and Pro Max models. iPhone 13 and iPhone 13 mini are ideal for those who don't need professional quality camera features.

The 5.4 inches 13 mini version than usual is the replacement to the iPhone 12 mini model, while the 6.1 inches iPhone 13 replaces the (2020) 12 model. The two new iPhone 13 models are almost identical in design to iPhone 12 models, with flat edges, an aerospace-grade aluminum case, a glass back, and a slight increase in thickness.

The screen of the new iPhone is brighter than its predecessor. The battery life is longer. What's more, the Apple Company has updated an all-around incredible camera with refreshed and advanced sensors and PC photography highlights that can influence you to have the skill and attitude of a star even on a non-professional telephone device.

It was said that at the beginning of 2021 no matter a single iPhone will have a portable modification that relies completely on wireless charging, but that won't happen this year. Apple plans to depend on lightning in the nearest time without switching to USB-C but would not a porting the modification till MagSafe accessories become available. The appearance of portable iPhones may start in 2022.

IPHONE 13 USER GUIDE

You'd expect a smaller level on the front of the iPhone, Apple can reduce Face ID hardware. The smaller size will be the only significant visual change to the front of the iPhone. Nevertheless, people anticipated that Apple would familiarize the 120Hz screen rejuvenate rates for Pro iPhone models, conveying the first iPad Pro ProMotion technology to the iPhone.

Apple was considering adding a screen fingerprint sensor to iPhone 13 models that were applied in the adding to Face ID as an alternative method of biometric verification, and it is a feature that is stored or stored for future updates because it did not stop on iPhone 13. The iPhone 13 will rely on Face ID only.

As for the improved camera technology, you think you see a bigger sensor at the iPhone 12 Pro Max that extends to different models, as well as the Sensor-Shift Steadiness has the best autofocus with balance. With the camera modifications, come design changes to the iPhone 13 models. It is going to be so obvious on the new iPhone 13 Pro, which gets an ample greater Pro Max-sized camera display, while the iPhone 12 mini and 12 come with a diagonal camera setting.

IPHONE 13 USER GUIDE

Chapter 1

WHAT IS THE DIFFERENCE BETWEEN THE IPHONE 13

Features of iPhone 13

Design

Apple will continue to offer four iPhones in 2021; without resizing, you will imagine the 5.4-inch iPhone mini, the 6.1-inch inexpensive iPhone, the 6.1-inch Pro model, and the 6.7-inch Pro Max model; some people said that the iPhone 12 mini sales weren't enough, and Apple is still developing the iPhone 13 mini, even if it can account for only 10% of the total production of the iPhone 13.

No significant design changes are expected in the 2021 iPhone models, as the 2020 iPhone models make an important update in technology and modification.

IPHONE 13 USER GUIDE

Apple engineers believe that the 2021 iPhones are the "S" type of the 2020 models. Apple had used the "S" nomenclature in the past for ages when iPhone models introduced smaller upgrades.

The novel iPhones come with similar dimensions as the iPhone 12 models, and the thickness will be anticipated to upsurge by 0.26 mm. The iPhone 13 models will continue likewise to be a little thicker owing to the bigger batteries.

The normal model of the iPhone 13 could see the lenses diagonally rather than vertically by installing a dual-lens camera. This is not completely clear what the benefits it will have in place of the gadget has determined to give similar wide and ultra-wide lenses, and it has a lot of modifications in stock such as optical image maintenance for wide lenses that may require relocation.

The iPhone 13 Pro model comes in 0.2 mm thick, probably to accommodate a bigger battery, that has an ample bigger rear camera, permitting to rendering images. You may need a larger camera unit to stabilize the sensor shift, and it is the identical dimension as the rear camera unit from the iPhone 12 Pro Max.

IPHONE 13 USER GUIDE

It comes with a bigger camera size than the iPhone 13 Pro; it is shown in the pictures of the iPhone 12 pro in the case of the iPhone 13 Pro, which has been conspicuous with a bigger camera. iPhone 13 models can likewise have strong magnets on the inside and an unlikely matte color on the outside. And it has been said that future iPhones will be substituted by a matte black color that can substitute the gray space with a novel stainless-steel cover that decreases stains and fingerprints.

Likewise, Apple is obviously seriously thinking of colors like an orange, bronze-like color, nevertheless, it will not be the case and is available in the pink shade of the standard iPhone 13.

Color

Both iPhone 13 and iPhone 13 mini colors come in pink, blue, midnight (black), starlight (white), and product (RED). Non-slip models usually come in more fun color options, and that's true this year.

Both iPhone 13 Pro and iPhone 13 Pro Max has four colors: they are gold, black, silver, and Sierra blue. The Sierra Blue is novel in this current year, and it appears lighter more than the Pacific Blue which was in the iPhone 12 Pro and iPhone 12 Pro Max.

Display

The iPhone 13 series will be accessible in three different screen size selections, just like last year. The iPhone 13 mini retains the same 5.4-inch screen, though the iPhone 13 and iPhone 13 Pro both come in at 6.1 inches. The iPhone 13 Pro Max comes in a 6.7-inch show as that of the iPhone 12 Pro Max.

All four phones come with Super Retina XDR OLED technology. The non-Pro models will be up to 800 nits of brightness, and the iPhone 13 Pro max and iPhone 13 Pro can dial up to 1000 nits. Apple also says the latest show technology is more efficient.

Apple says that the iPhone 13 Pro and iPhone 13 Pro Max will feature a ProMotion show. This means that two phones can lock their screens up to 120 Hz or 10 GHz. This range will help make the content you are viewing more sensitive as iOS adapts the refresh rate to whatever you are doing.

Apple is also working with developers to enhance their applications and games for 120Hz applications.

Cameras and cinematic mode video

Apple has added a major camera upgrade for the iPhone 13. Non-professional models have a new 12-megapixel ultra-wide sensor that delivers 47% more light. The 12-megapixel main wide-angle camera has optical image stabilization, a feature previously limited to the iPhone 12 Pro Max. This will help iPhone 13 and iPhone 13 mini to take beautiful photos in multiple lighting conditions.

You also get a photographic style with each new iPhone that gives your personal preference for your photos in real-time. It's more than just a filter; it uses the correct corrections for different scenes to produce the best photos.

The four iPhone 13 models come with a cinematic mode that gives the best superior cinematography to the videos with Dolby Vision HDR. A requirement of the popularity of cinematic mode is its ability to dynamically change the focus and follow the subject of your video. If the subject is moving, the focus will also move. If a person is avoiding, the focus shifts to where the subject is looking. Just tap where you want to focus and then tap again to lock in focus.

The iPhone 13 Pro and iPhone 13 Pro Max will similarly shoot videos in ProRes format, which opens up plenty of professional-quality editing capabilities. Note, however, that the 128 GB version of the iPhone 13 Pro and Pro Max is covered by 1080P for ProRes video; Superior storage options are supported by 4K ProRes.

iPhone 13 spec

The new iPhone 13 is the A15 Bionic system on the chip, Apple's most improved chip forever. It is built on a 5 nm process with 6 cores. Two of them are new high-power cores, while the other four are more efficient than before. Also, with a neural engine advancement, the iPhone 13 comes with powerful machine-learning apps like iOS 15's new live text feature and speech recognition on the Siri device.

The new iPhone 13 mini and iPhone 13, have the quad-core GPU, which is perfect for graphically gaming and photography. In fact, Apple claims that the GPU is 3% faster than most you will find on an Android phone.

The iPhone 13 Pro and iPhone 13 Pro Max will get an even bigger GPU upgrade with 5 cores. Apple claims that the Pro models are equipped with the most powerful mobile GPU, which is probably 50% faster than the competitor. With the new possibilities of photography and cinematography, it can be a huge stimulus for creative productivity.

Small size

The iPhone 13 is anticipated to appear at a lower level, but numerous trustworthy sources indicate that the level size will decrease in 2021

The TrueDepth camera can be more superficial as the receiver at the upper part of the screen moves to the edge of the case. The data initially advocated that the hole would be more superficial than small, but it was later mimicked that it would be reduced in width moderately than height.

The "reduced" size on the iPhone 13 models will be made a little recodification camera module that combines the Rx, Tx, and Flood Illuminator to reduce the dimension, and little VSCEL chips.

It has been said that achieving a lower level than the iPhone 13 models would be the result of a "more closely integrated version of the existing structured lighting system."

The Face 13 transmitter of the iPhone 13 is designed with plastic in its place of glass, which becomes imaginable to enhanced coating technologies, and it is unclear whether this will benefit the owners when it simply means a decrease in items of manufacturing.

Lesser sizes will be viewed in the leaked screen panels, which is likely for iPhone 13 models. The panels have smaller holes because the headphone speaker is moved to the top frame.

The images of a 3D-printed model of the iPhone 13 were shared, displaying a lesser scratch, a front-facing camera on the left, and a removable headphone jack. Likened to the iPhone 12 Pro, the mock-up advocates that the iPhone 13 Pro is 5.35 mm tall, as opposed to 5.30 mm, and 26.80 mm wide, below 34.83 mm.

Portal design

Apple's long-held success was an iPhone that lacked outside ports and tab for clean, streamlined gadgets, and a hat Apple will present a portable iPhone in 2021 without a Lightning port, and that may not happen.

Apple has decided to remove the Lightning port on some iPhone models is the best for wireless charging, but Apple would remain to the use of the Lightning connector for the 2021 iPhone.

IPHONE 13 USER GUIDE

The handy iPhone is charged through the use of Qi-based wireless charging and MagSafe accessories,

USB-C is not

Apple is not going to switch from iPhone lightning to USB-C, Apple does not want to switch to USB-C since it is an exposed standard with little waterproof than Lightning.

The Lightning, Apple will be adjusting the superiority of Lightning cables with the accessories that were designed for iPhone (MFi) software, and this is not conceivable with USB-C.

It is expected that the iPhone 13 and forthcoming iPhones will still be using the Lightning ports till MagSafe accessories become widespread sufficiently to enable the creation of a portable iPhone design.

No Touch ID

During the development of the iPhone 13, Apple tested a touch screen reader of the Touch ID screen, which will significantly change the biometric information that Apple has used since the release of the iPhone X in flagship devices, enabling dual biometric functionality.

Unfortunately, Touch ID at the bottom of the screen did not "stop" the iPhone 13 models, and Apple will continue to focus only on Face ID. The iPhone 13 will not have a Touch ID fingerprint sensor.

Apple is focusing on Face ID and plans to insert Face ID under the screen to design all screens of the future iPhone rather than using Touch ID on the next flagship device.

If Apple used Touch ID on the screen, the fingerprint sensor would be available for the dual biometric system in addition to Face ID

120Hz ProMotion display

Prior to the release of the iPhone 12, there were some innovations that recommended that high-end iPhone 12 models might have a 120Hz ProMotion display, and after the news cycle, it was revealed that this specification was held back till 2021 because of battery issues.

To do this, the 120Hz update speed on the iPhone, Apple should use LTPO show technology, which happened in 2021. Two models of the 2021 iPhone can use OLED show with low-power LTPO rear technology, paving the way for 120Hz upgrade speeds. LTPO gives a lot of energy-efficient background that switches on and off personal pixels on the screen but gives the best battery life. The two models with LTPO will be the Pro models, which are said to receive ProMotion technology.

Samsung will provide an LTPO OLED show that can permit Apple to deliver 120Hz upgrade speeds on 2021 iPhone models, while BOE and LG Show will likewise supply Apple screens for use with the novel iPhones. Samsung started the production of the LTPO OLED show for Apple in May 2021. Apple will apply Y-OCTA show technology that will permit you to sample a touch screen layout on an OLED panel that has no need for a separate layer. The results will be in a thinner show that is likewise more accessible.

Apple providers Samsung and LG Show, are anticipated to transform the section of their manufacturing to produce little–energy to the LTPO OLED panels for the iPhone 13. The comprehensive adaptation of the manufacturing from LTPS screens to LTPO is projected to be accomplished in the first half of 2021.

Constantly on show

The iPhone 13 may have a screen similar to the Apple Watch, this will likely allow some data such as time to be displayed on the locked screen of the iPhone even when it is turned off.

There is also evidence that the iPhone 13 will have a constant on-screen that looks like a low-lock screen, likewise with a clock and battery charge level. Messages appear, but the screen does not fully illuminate.

A15 processor

Apple will use the 5 nm + A15 chip for 2021 iPhones, and the chip will be produced by TSMC. And the novel chip design constantly brings enhancements in rapidity and effectiveness.

TSMC began manufacturing the A15 chip in May. And Apple has requested about 100 million A15 chips for novel iPhones in anticipation of the increase in requests to upgrade models.

WiFi 6E and 5G chip

Qualcomm, in February 2020, introduced the novel Snapdragon X60 third-generation 5G modem that is going to be used in 2021 iPhones. Forthcoming iPhones will equally use X65 and X70 Qualcomm modem chips.

Qualcomm's Snapdragon X60 is designed on a 5-nanometer development, which offers higher energy effectiveness for small footprints than the X55 chip used in the iPhone 12 models. Equally, it provides a combination of 5G data from mmWave and sub-6GHz bands to instantaneously improve speed and coverage.

Quicker mmWave 5G technology will be expanded to other nations, and the number of Apple mmWave iPhones will grow meaningfully in 2021. Over 50% of iPhone 13 models are anticipated to back by mmWave technology, while Apple is preparing to order more mmWave antennas

IPHONE 13 USER GUIDE

It is expected that Apple will remain to reduce iPhone costs in 2021 is a good fund to the costs of 5G chip technology, accepting a soft battery board modification that will reduce costs by 30 to 40% likened to the cost of the iPhone 12 board.

It is believed that the iPhone 13 models will back WiFi 6E, which gives WiFi 6 specifications and capabilities in the 6GHz band. WiFi 6 offers higher presentation, lesser latency, and quicker data speeds, with the extra range of WiFi 6E givers amplified bandwidth above present 2.4 and 5GHz WiFi bands. It was also predicted that Apple could support WiFi 6E as part of the 2021 iPhone.

Larger batteries

The iPhone 13 models equally come with bigger batteries that can lead to lasting battery life. Apple has created a novel space-saving design, like incorporating a SIM card into the motherboard and lowering the thickness of the former optical module to give more space for the battery.

The iPhone 13 Pro Max has a 4352 mAh battery, from 3687 mAh in the iPhone 12 Pro Max, and the iPhone 13 and iPhone 13 Pro have a 3095 mAh battery, from the iPhone 12 2828 mAh. The iPhone 13 mini is anticipated to add a 2406 mAh battery, equated to the 2227 mAh battery in the iPhone 12 mini.

The A15 chip introduces additional enhancements in efficiency

iPhone 13 Pro models have 15-20% reducing power consumption to replace battery life with an in height to rejuvenate the rate

iPhone 13 Mini battery lasts more than an hour on the iPhone 12 mini, and the iPhone 13 Pro Max will get 18-20% more battery life. The iPhone 13 and 13 Pro will have 10% larger batteries, but the battery life of the iPhone 13 Pro may be

IPHONE 13 USER GUIDE

followed by the battery life of the iPhone 13 due to the 120Hz ProMotion display.

Apple has added bigger batteries in all four models of the iPhone 13 this year. While the company has not announced full capabilities, it said it would give us rough estimates of how long the iPhone 13 options will last.

For the iPhone 13 mini, Apple said the phone will last an hour and a half longer than the iPhone 12 mini. The iPhone 13 seems to last two and a half hours longer than the iPhone 12. The iPhone 13 Pro will take an hour and a half longer to work than the iPhone 12 Pro, while the iPhone 13 Pro Max will take up to two and a half hours. More than an hour before its predecessor. Apple notes that the iPhone 13 Pro Max is the most durable iPhone ever.

25W power adapter

iPhone 13 models can accelerate charging speed by supporting a 25W power converter. The iPhone 12 model gives quick charging, and 20W offers maximum charging power.

Apple will likewise give a novel 25-watt adapter that will come into sale with newer iPhone models.

1 TB concentrated storing

The iPhone 13 model comes with 1TB of storing space that will come up to 512GB in iPhone 12 models.

iPhone 13 and iOS 15

The best features of iOS 15 include FaceTime updates that allow you to watch videos and share content with others, a new focus mode for managing messages, and improved Maps, Messages, Wallet, Weather, and other built-in apps.

IPHONE 13 USER GUIDE

iPhone 13 Perspective

Much of the focus of the iPhone 13 is on the cameras and the A15 Bionic chip. These are all notable updates. The hole is smaller (eventually), and the photography capabilities are certainly impressive. The iPhone 13 seems to offer a lot more than its predecessor.

Pro models have a 120Hz ProMotion display, which will help iOS 15 look particularly sleek. All four models of sensory shift stabilization are fantastic, as well as extremely extensive improvements. And do not forget about battery upgrades.

Tips and tricks of iPhone 13

Present version

Apple released eight beta versions of iOS 15 to inventors and seven beta versions to community beta samples.

Originator Beta Five has announced many improvements like a novel weather app symbol, a redesigned camera and voice acknowledgment symbol in the Control Center, TestFlight info in the App Store, a warning that iPhone can be found even when turned off, and new release screens describing apps such as maps, Homes, and Reminders.

IPHONE 13 USER GUIDE

Developer Beta Six has made a number of changes, including a major overhaul of the Safari interface that provides two options for the tab bar location and restores previous changes, Safari bookmarks preview, read list and history view in the bottom tab, Shortcuts, return timers, AutoPlay and the removal, that has been overdue and will no longer be accessible to the public with the launch of the iPadOS 15.

Redesign messages

IOS 15 introduces fully recycled messages. Messages now show contact pictures of people and bigger symbols for apps to be easier to be recognized.

It has a brand novel customized message summarization that collects non-important messages organized to be provided at a more suitable time, such as evening or morning. Messages in the resume are prioritized by means of gadget cleverness, which analyzes operator relations with apps and ranks the greatest significant and important messages high. Critical messages and time thoughtful messages will be provided instantaneously.

Now the ability to temporarily mute threads from any app or message strand and iOS 15 will offer you to turn off strand if it is abnormally active and you will not be part of it.

It has a novel messaging API for inventors that allows you to send time-sensitive messages and get a novel view of messages received from people.

Focus

IOS 15 includes a novel device to aid operators to lessen attention span named Focus. The focus will filter messages with the home screen pages depending on the desire of the user wants to focus on a specific time, even when sending emergency messages. If the user focus is presently obstructing incoming messages, the status will automatically appear in different persons' messages, indicating that the user is presently unavailable; maybe it is crucial.

The iOS can automatically give users attention to the usage of the gadget intelligence to decide the people, but the applications will inform them at an assured period. Focusing propositions are created on the context of the users, like the working periods, individual time, and going to bed. Users will likewise set up custom focus to show only assured messages hours and home screen pages, choose permit delays at the very vital messages but the application and create automatic responses to messages while focusing.

If the user starts the focus, it syncs to the Apple gadget. Inventors will put the novel status API for third-party messaging applications to show the focus status.

In the center of attention

Spotlight will now henceforth you need to apply the apply aid of intelligence to look for objects, photos by location, scenes, people, or objects, and it will be through the novel live text specification, Spotlight will find text and handwriting in the photos.

Spotlight similarly backs web image search but meaningfully rich results for musicians, actors, movies, and TV shows. Improved contact card results display current chat's location, and shared photo's location when shared via Find My.

Nevertheless, similarly is thinkable to have contact to Spotlight via a locked screen but speedily connect to applications through the App Store deprived of exiting the Spotlight. For the activities that back app clips, it has a Spotlight action tab at the Maps.

IPHONE 13 USER GUIDE

FaceTime

In iOS 15, FaceTime gives space audio so that video call sound like when they are coming from where an individual is on the screen. It has novel microphone modes to detach the user noise at the background noise or optionally make known background noise in a wide range of modes.

FaceTime currently has portrait mode for video calls, so users can blur their wallpaper but focus on themselves but spoil the alerts, so it's clearer if you're talking mute. It likewise has a novel network view for group FaceTime calls, permitting members to view numerous faces at the instance and an optical zoom regulator for the rear camera.

SharePlay

SharePlay is a novel specification that permits operators to share familiarities at the time of FaceTime calls, comprising of the media such as Apple Music songs, or movies TV shows. The Media harmonized every member but is shared playback, so other people in the SharePlay session will play, skip or pause content, and do well to include the shared queue. SharePlay likewise offers users the capability to share their screens to see the applications with a FaceTime call.

Third-party apps like Disney+, ESPN+, HBO Max, Hulu, MasterClass, Paramount+, Pluto TV, TikTok, Twitch, and others are required to share SharePlay.

SharePlay spreads to iPad, Mac, iPhone, and Apple TV, so users will equally see movies or shows while connecting to a larger screen with FaceTime. Intelligent voice control enthusiastically and automatically regulates the audio that you will listen to the friends even if it is shared content is playing out loud. SharePlay also features in-app messaging controls.

FaceTime Links

Users can create a FaceTime call connect and share it via messaging, calendar, email, or third-party applications.

FaceTime links will be opened on Apple devices to use the FaceTime app, and they will likewise be opened via a web browser, for the first time on FaceTime Android and Windows. FaceTime calls on the Internet remain fully encrypted to ensure privacy.

Photos

The Photos app features a significant upgrade to Memories, featuring a novel design, combination with Apple Music, a lot of collaborating interfaces, and a memory layout.

Memories offer songs created on the Apple Music listening history that are synced with photos and videos for a more customized experience. Users will personalize memories by flipping via memory mixes, allowing you to learn varying songs at varying tempos in addition to the atmospheres.

It has 12 kinds of memory that create the mood by explaining separately videos and photos with the accurate consistent look of contrast and color correction. It also novel kinds of memory, comprising of extra international holidays, child-centered memories, was in invoke for some time, but pet memories, comprising of the ability to identify the cats and dogs.

This is similarly imaginable to see but edit each of the contents of the memory very fast, while the Watch in Subsequent segment offers correlated memories.

The individual identification specifications improve personality recognition, and it makes it simple to correct naming errors in individual albums. It has a specification less selection that permits the pictures to know that you desire to display little

IPHONE 13 USER GUIDE

of a precise place, date, holiday, or individual featured in the Favorite Photos, Photos Memories, Library button, and Widget.

At iOS 15, Pictures contains a better-off information window to see data on the pictures, like shutter speed and camera, lens, file size, or the person that sent the photo messages to you. You will likewise change the location or date taken, including captions, with the view items identified by Visual Look Up.

Photo Image Chooser, comprising of the Messages application, present-day permits you to choose pictures in a detailed order of sharing. The Third-party application will similarly give simpler workflows if you grant access to precise content in a picture's library.

It is very fast, with initial syncing of iCloud Pictures to a novel gadget is quicker in iOS 15.

Maps

Maps have a global interactive interpretation and meaningfully heightened details in the novel 3D city view. Commercial areas, Neighborhoods, heights, buildings, and a lot are present-day displayed in detail, with novel road hue and labels, personalized landmarks, and a novel "moonlight" mode for the night.

The Maps application likewise gives a novel 3D driving understanding in the city with road specifics like the curves, medians, bike pedestrians, and lanes crossings running on the iPhone even with Apple CarPlay.

Transportation direction-finding has been modified, and travelers will currently and more easily locate the close by stations and attach their preferred lines. Maps automatically move with the chosen transit route and notify users if it is the time to arrive.

IPHONE 13 USER GUIDE

It has a novel view of movement guidelines that use augmented reality. Users will basically grab their map application and the iPhone to generate a correct position using the camera to give detailed driving directions.

Remanufactured location cards perform it simply to look also interrelate with info concern your business, location, with physical characteristics. With a novel guide-home that contains data concerning the novel places with editorial.

When searching for a new location, there are new options with results filtering criteria such as cooking or opening time. Maps also automatically update search results as you navigate, and the most commonly used settings are located in one, simpler location.

Wallet

The Wallet app backs extra kinds of keys in iOS 15, like the office, home, hotel, or corporate number key cards.

The Wallet applications likewise expand backing for car keys, present-day by means of Ultra-Wideband to unlock, lock and start the car deprived of removing the iPhone from the pocket. Ultra-Wideband likewise gives delivers accurate spatial awareness, it then means iOS can preclude you from locking the car, and the iPhone is inside, or start the car if the iPhone is outside.

The wallet now supports remote access control, allowing you to lock or unlock your car, shout your horn, warm up your car or open your trunk.

Later, from 2021, users in US member states can be enabled to include their driver's license or ID card in the Wallet app. Apple says that the Transport Security Administration is working to activate airport security checkpoints, as customers will primarily use a digital ID card in their wallet.

IPHONE 13 USER GUIDE

The Wallet app will also automatically archive expired check-in cards and event tickets. Safari likewise backs including of numerous passes to the wallet in a single action, in its place of manually including one pass at a time.

Cover

iOS 15 is coming with a whole novel modification to Safari. Taken controls at the bottom of the screen so that one hand can more easily reach them.

This is a novel, compact button bar that flows at the down of the screen so that the users will effortlessly switch among buttons, and it also contains a smart search field. Click on the groups permits users to store their buttons to a file and sync to iPhone, iPad, and Mac. Additionally, there is a network view of the new tab overview.

Users can simply drop the webpage to update it, and now there is voice search support. Safari also acquires a personalized homepage and mobile web extensions for the first time.

iOS 15 features Safari's new privacy features, including intelligent tracking prevention, which prevents trackers from profiling your IP address, and Safari automatically updates sites that are known to support HTTPS from unprotected HTTP.

Messages

Content sent by you will be automatically displayed in messages in the new "Share" app. Shared with photos presented in Pictures Apple News, Apple Podcasts, Safari, Apple Music, and the Apple TV application. Users can capture content shared with them so that it grows in search of shared messages with you and in conversation details

The Collections of photos sent in messages will now display as a collage or folding stack, according to how many have been sent. It is also now possible to find images shared through messages using the contact's name.

Messaging likewise receives regional enhancements in iOS 15, like unsolicited SMS filtering in Brazil and messaging settings in India and China, allowing users to turn off messaging for the type of messaging they choose.

Weather

The weather information is completely restructured in iOS 15. The weather application present day has many graphical shows for weather information, a complete screen map, and a dynamic layout that varies depending on external conditions.

Apple has restructured an animated weather application background to more correctly reflect the present state of the sun and rainfall conditions. There are also messages that highpoint if snow or rain stops and starts.

Health

At the iOS 15, the health application has a novel allocation button that will permit users to portion their selected health information with the caregivers or family. Lab results are enhanced with descriptions, indicators, and the ability to attach results for quick access. Health will currently notice the trend that will draw consumers' responsiveness to significantly modified individual health indicators.

The new Health applications similarly include Walking Steadiness as a novel metric to aid control fall risk. The test result of COVID-19 immunization can be stored in healthcare data via the healthcare provider QR code. Blood glucose readings now show levels during sleep and exercise and have interactive charts.

IPHONE 13 USER GUIDE

Chapter 2

FEATURES OF IOS 15

Facetime

To reduce background noise, use voice isolation mode, and portrait mode to center the visual attention on you. People in your facetime conversations group are displayed in a grid format in the same-size tiles, with the speaker highlighted automatically. You can now use a new web URL to invite anyone to join a facetime call.

Messages

Messages with several photographs now show as a college or group of images that you can swipe through for convenient viewing.

IPHONE 13 USER GUIDE

Memoji

Your appearance and style will be reflected in different ways. For your memoji stickers, there are new costumes, additional headgear, and improved personalization possibilities.

Focus

Focus allows you to automatically filter alerts based on your current activity. Choose a focus recommendation from the drop-down menu, such as work, sleep, or personal, and then choose the alerts you want to receive during certain hours. When someone outside of your authorized notifications attempts to contact you, your focus status displays in messages, letting them know you're unavailable.

Set up your focus.

Do not disturb, personal, work, sleep, and new focus are the options from top to bottom.

Notification

Notifications have a new appearance, with bigger app icons and contact pictures to make them simpler to recognize. You may also get a daily notification summary containing a selection of notifications depending on a schedule you establish. A summary of how to schedule notifications may be found here.

Map

Elevation, landmarks, trees, crosswalks, and more are all included on maps for cities like San Francisco and New York. When approaching complicated interchanges, new driving features provide road information like turn lanes and bike lanes, as well as street-level viewpoints. Public transportation capabilities have been updated to highlight local stations and transit schedules, as well as augmented reality walking guidance.

Safari

The downward part of the screen tab bar makes it easier to access and open your tabs. Tab groups make it easier to arrange tabs and switch between them.

Live text

You can now send an email, make a phone call, or copy and paste with a single tap in live text because it recognizes text in photographs and within the iPhone camera frame.

A sign with text and a phone number is shown in the shot. After selecting the phone number, the following choices appear on the screen: call, text, facetime, facetime audio, add to contacts, and copy are all available options.

Visual lookup

Look up recognizes things in your photographs, such as well-known locations, plants, literature, and art, and then search the web for more images and information (the U.S. only).

Memories

The memory and other editing options in the photo, help you customize the appearance and feel of your memories. You may also use apple music to play selected music in the backdrop of a memory.

Reminders

Add tags to your reminders as you make them, such as #errands or #homework. Create smart lists to arrange reminders automatically based on tags, dates, times, places, priorities, and more.

Weather

Animated backdrops and graphical representations of precipitation, air quality, and temperature maps are part of the new design.

The current location, favorite locations, and overlay menu buttons are located in the upper-right corner, from top to bottom. When you choose the overlay button, you'll see options for altering the screen display to indicate temperature, precipitation, or air quality; tap 'Done' in the top-left corner.

Senders attempting to obtain information about your postal activity, track your internet activity, or establish your location are protected by privacy mail privacy protection. It also keeps senders from knowing whether or not you've opened their email.

iCloud+

iCloud+ includes everything that iCloud currently has with added premium features, including iCloud private relay (beta), which secures your online privacy.

For only the applications you desire, enlarge text or use boldface, improve contrast, reverse colors, and more. With voiceover, you may look at people, objects, text, and tables in photos in greater depth. By playing calming sounds in the background, you may mask undesirable ambient or external noise.

Note that based on your iPhone model, location, language, and carrier, new features and applications may differ.

Chapter 3

SET UP IPHONE 13

Understanding your options

You can easily set up your smartphone in one of the three ways: begin afresh, restore from another iPhone, or import the content from a non-Apple device. Here are the details of each option:

Set up as a new device: This simply means beginning everything from scratch. This method is for individuals that have never used a smartphone or any online service that's owned by Apple before.

Restore from a previous Apple device: You can restore online via iCloud or with a USB for those using macOS Catalina or later. This option is for individuals that have owned an iOS device before and wish to move to a new one while keeping the content of the older device.

Import from Blackberry, Android, or Windows smartphone: There's an app that Apple has in Google PlayStore to make migrating from Android to iPhone an easy step. This app allows you to easily move lots of data to your new iPhone. This method is designed for individuals that are switching to an iPhone or iPad from a different mobile platform.

IPHONE 13 USER GUIDE

Once you switch on your new iPhone for the first time, you will see "Hello" in different languages. This is the same whether you're starting from scratch or restoring your device from another iPhone, or switching from an Android device. Here are the steps to follow to set up your iPhone 13:

- Touch slide to set up, and get started by sliding your finger across the screen.

Choose your preferred language

Choose your preferred region or country

Choose a WiFi network. If there's no Wi-Fi network within range, then you can make the set-up at a later date. In such an instance, select "Cellular."

Here you can opt for Automatic Setup to set up your device with the same passcode and settings as a different iPhone.

If you'd rather set up your device manually, you can continue with the steps below.

Click on "Continue" after reading Apple's Data and Privacy information,

Click on "Enable Location Services." if you're not willing to enable location services now, then tap "Skip Location Services." Certain location services like Maps can be enabled.

How to wake and unlock iPhone 13

How to Wake and Unlock your iPhone 13 series

In order to save power, your iPhone 13 will usually turn off the display (light) and will enter sleep mode when you are not currently using it.

IPHONE 13 USER GUIDE

You can use any of the below methods to wake and unlock your device:

- Use the Side button: The Side button is found on the right side of your device. Click on the Side button to wake up your iPhone.
- If you don't like to use the Side buttons, you can also wake up your iPhone by merely raising it up. This feature is called the raise to wake feature. This feature can be turned off by navigating to Settings and then selecting Display and Brightness to turn OFF.
- You can also wake up your iPhone 13 [all models of iPhone 13] with a mild tap on the screen.

How to unlock iPhone with Face ID

The Face ID offers a secured means of unlocking your device, purchasing and making payments online, and signing in to some third-party apps by merely placing your face on the appropriate side of your iPhone.

However, you will not be able to use the Face ID just like that if you don't set it up either now or during the initial configuration of your iPhone. To set up the Face ID feature now, follow the steps below:

Set up a passcode for your iPhone 13: Navigate to the Settings app on your iPhone, and click on Face ID and Passcode.

You will need to set up a passcode before you can use the Face ID feature. Setting up a passcode will help activate data protection for your device. Click on Turn Passcode ON or select Change passcode. Select passcode options if you want to see options for creating a passcode. Choose a Custom alphanumeric code or custom numeric code as they

IPHONE 13 USER GUIDE

tend to be more secured because these two options let you customize your own passcode.

Set up the *Face ID* by

Navigate to your iPhone Settings⚙, select Face ID and Passcode, and tap on Set up Face ID. Follow the instructions prompted on the screen for a complete Set-Up.

You can add an extra face appearance that can complement the one you have already added. Be aware that this alternate face appearance will be able to unlock your device even when you are not present there. To do this, navigate to Settings⚙, select Face ID, and Passcode, and tap on "Set up an alternate face appearance." Follow the instructions shown on the screen to complete the setup.

For people with one or two physical limitations, especially blindness or poor sight, simply tap on Accessibility options when you are setting up Face ID. Here, you will be able to restrict your device from asking for your full eyes opened before it unlocks. To do this, navigate to Settings, tap on Accessibility and toggle the "Require Attention for Face ID" switch to OFF.

How to Turn Off Notification on your iPhone 13

This article gives information on how to turn off notifications on your iPhone, both temporarily and longer. It also helps you to know what happens when you've turned off the feature:

How to turn off notifications temporarily

Temporarily turning off notifications on your iPhone 13 is a very easy step. You can do this by simply enabling the Do Not Disturb feature on your device either by the Control Panel or Settings. The Do Not Disturb feature will turn off all the notifications that arrive on your phone, this includes the app notification, message and mail notifications, call notifications.

Steps to Turning off Notification on your iPhone

Step 1: Navigate your way to the Settings app, then tap on "Notifications."

Step 2: You can alter the notifications on your app here. Click on the app to adjust the notification as you desire.

Step 3: Select the option that best suits you for the notifications. Select disable "Allow Notifications" or alter how and where sounds, badges, and notifications appear.

Once you've made the selections you want, leave the page. When you want to get notifications again, follow the instruction to alter the settings again.

Change Notification Settings from an Incoming Notification

There's also an alternative to silencing notifications for specific apps with this method. If you received a notification from an app that you want to turn notifications off for, you can simply adjust the notification settings directly from the notification.

Here are the steps to getting this done:

Step 1: On the notification screen that pops up when you wake your iPhone, slowly slide a notification to the left. You should see a menu on the right side of the notification.

Step 2: Tap on "Manage."

Step 3: On the screen that pops up, click on "Deliver Quietly" or "Switch Off." if you'd rather opt for Deliver Quietly, you will still get the notification in your Notification Center, but this won't vibrate or make noise on your phone.

If you'd rather go for Turn Off, then your phone will switch off notifications totally until it's turned back. You can simply click on Settings to be redirected to the same settings screen that you used in the steps above.

Turning Off All Notifications on your iPhone 13

The easiest way to silence all notifications on your iPhone 13 is via the Do Not Disturb option which is detailed above. However, if you desire true quiet time, then you'll have to simply flip the switch to the left side of your iPhone 13.

Doing this automatically silences all notifications, and they'll remain silenced till you can move the switch back to the on position.

What Happens When you Switch Off Notification on Your iPhone

If you're bothered about the long steps required to switch off notifications on your iPhone, you need not be. Although turning this feature off by using either Do Not Disturb or one of the methods to change the notifications you receive for individual apps, you can always switch the feature back on anytime you're ready.

How to Pair Hearing Aids to iPhone 13

This article is an explanation of how to connect an iPhone-Compatible hearing aid with your iPhone. To use this type of hearing aid, you need to have an iPhone-compatible hearing aid and, of course, your iPhone 13.

Connecting Hearing Aids to your iPhone 13

After making sure that your iPhone and your hearing aid are compatible, the next steps are to begin the process of connecting the two devices together. This process only requires a couple of taps to accomplish. Here are the steps to follow:

Step 1: First, ensure that your Bluetooth on your iPhone 13 is turned on

Step 2: Navigate your way to Settings, then tap Accessibility, Hearing, and then choose "Hearing Devices."

Step 3: Open the battery doors on your hearing aid, and then close them. This puts your hearing aid into pairing mode while your iPhone is looking to establish a connection with it.

Step 4: Your hearing aid will pop up under the MFI Hearing Devices heading. When you see this pop-up, click on your device and choose a pair.

Step 5: It can take up to a whole minute for pairing to be completed for the two devices. Once you've paired the devices, you can start to use your hearing aid as you'd normally do

Back up iPhone with iCloud

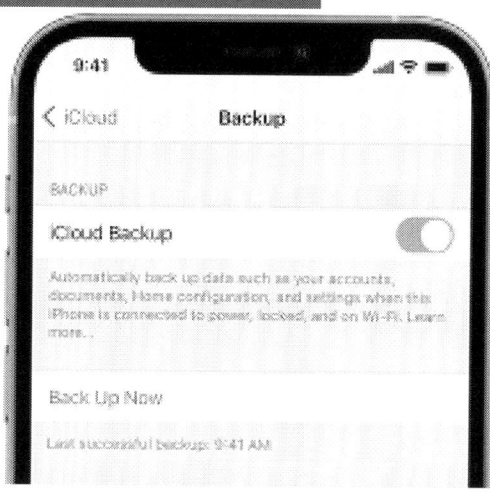

To access iCloud Backup, move to Settings > [your name] > iCloud > iCloud Backup.

Enable iCloud Backup.

iCloud backs up your iPhone automatically every day when iPhone is plugged to power, locked, and linked to Wi-Fi.

Nota Bene: On 5G-enabled models, your carrier can provide you the option of backing up your iPhone over your cellular network. Navigate to Settings > [your name] > iCloud > iCloud Backup, and toggle Backup Over Cellular on or off.

Click on Back Up now to activate manual Back-Up, move to Settings > [your name] > iCloud > Manage Storage > Backups to see your iCloud backups.

To remove a backup, press Delete Backup after selecting a backup from the list.

Back-Up with Windows PC

A cable is required to connect your iPhone to your computer.

Click the iPhone icon at the upper left of the iTunes window on your PC's iTunes app.

Press Summary.

Now, touch Back-Up (under Backups).

Encrypt your backups by touching Encrypt local backup, inputting a passcode, and then hitting Set Password.

To see the backups that are saved on your computer, move to Edit > Preferences > Devices. Encrypted backups are indicated by a lock symbol in the backup list.

Automatically update your iPhone

If you can't activate automatic updates when you first set up your iPhone, take these steps:

To get automatic updates, move to Settings > General > click Software Update > Automatic Updates.

Switch on Download iOS and Install iOS Updates.

Manually update your iPhone

You can install software updates at any time.

Hit Settings > click General > Software Update.

The screen indicates the version of iOS that is presently installed and if an update is available.

To disable automatic updates, move to Settings > General > Software Update > Automatic Updates and uncheck the box next to Automatic Updates.

Physical SIM installation

To eject the SIM tray, insert a paper clip or SIM eject tool into the tiny hole of the SIM tray and press it into the iPhone.

The shape and orientation of the SIM tray are determined by the iPhone model and the nation or area in which you live.

Pull the tray out of your iPhone

Insert the SIM card into the tray. The proper alignment is determined by the angled corner.

On the iPhone, a SIM card is placed into the tray; the angled corner is at the top left.

Put the tray back in iPhone.

If you have already configured a PIN on the SIM, enter it carefully when asked.

WARNING: Never attempt SIM PIN guessing. A wrong guess will permanently lock your SIM, preventing you from making phone calls or using cellular data via your carrier until you get a replacement SIM.

Set Up eSIM

The iPhone XS, iPhone XR, and subsequent models can digitally store an eSIM supplied by your carrier.

Enter Settings > Cellular and select Add Cellular Plan.

Take one of the following actions:

Utilize a QR code given by your carrier to create a new plan: Adjust the iPhone so that the QR code displays in the frame, or manually input the information. Your carrier may need you to input a confirmation code.

Install a given cellular plan: If your carrier notifies you that you've been allocated a plan, select Carrier Cellular Plan Ready to Be Installed.

Transferring a SIM card from an old iPhone to a new iPhone: From the list, choose your phone number.

If you can't see your mobile number, ensure that both iPhone gadgets are logged in using the same Apple ID. Transferring a SIM card to an eSIM is not available on all carriers or cellular contracts.

Press Add Cellular Plan.

If the plan is your second line, use the on-screen instructions to configure how the plans should operate together.

IPHONE 13 USER GUIDE

Change physical SIM card to eSIM

On iPhone XS, iPhone XR, and later models, you can change a physical SIM to an eSIM if your carrier enables it.

Navigate to Settings > Cellular and choose a cellular plan that requires an actual SIM card.

Convert to eSIM by pressing Convert, then following the on-screen instructions.

Explore 5G

iPhone 13 and iPhone 12 devices are compatible with some carriers' 5G cellular networks.

What you need

- iPhone 13 or iPhone 12.
- A 5G-compatible carrier.
- Cellular plan for 5G
- Recognize the 5G status bar icons.

When you're in an area covered by your carrier's 5G network and your 5G cellular plan is active, your iPhone's status bar will display a 5G icon:

5G: The 5G network of your carrier is available, and your iPhone may connect to the Internet through this network.

5G+ and 5G UW: The higher-frequency version of 5G offered by your carrier is available, or your iPhone is currently connected to that network. Not accessible in all places.

5G UC

The 5G UC network of your carrier is accessible. This includes a higher frequency of your carrier's 5G version. Not accessible in all places.

Discover more 5G options

On iPhone, the default settings for 5G are designed for battery life and data consumption, depending on your data plan. In certain applications, you can choose when to utilize 5G and how much data to use.

Navigate to Settings > Cellular > Cellular Data Settings to see these options. If you're using Dual SIM, move to Settings > Cellular and choose the number for which you'd want to modify the settings.

Data and Voice

You can customize how your iPhone connects to the 5G network, which impacts battery life.

Enables Smart Data mode with 5G Auto. When 5G speeds are not much faster than LTE, your iPhone automatically switches to LTE to save battery life.

5G Enabled: Always connects to the 5G network when it is available. This can result in battery life reduction.

LTE: Utilizes the LTE network exclusively, even when 5G is available.

Data Mode

Allow for More Data on 5G: Enables capabilities that need more data for applications and system activities. These enhancements include improved FaceTime quality, Apple TV high-definition content, Apple Music songs and videos, upgrades for iOS through cellular, and automated iCloud backups.

When the Wi-Fi connection is sluggish or insecure on a network you frequent regularly, this option allows your iPhone to automatically switch to 5G.

IPHONE 13 USER GUIDE

To toggle this function on or off for a particular network, enter Settings > Wi-Fi. Select the info icon next to the Wi-Fi network and then hit Use Wi-Fi for the Internet.

Furthermore, this option enables third-party applications to use additional cellular data for better experiences.

Depending on your carrier, this is the default option for certain unlimited-data plans.

This option consumes a greater amount of cellular data.

Standard: Activates automatic updates and background operations on cellular, and uses standard video and FaceTime quality settings. This is the default setting in most cases.

Low data mode: Reduces Wi-Fi and cellular data use by stopping automated updates and background activities.

Link iPhone to Wi-Fi network

Navigate to Settings > Wi-Fi and enable Wi-Fi.

Select from the following:

A network: If prompted, provide the password.

Other: Connect to a hidden network. Input the name of the hidden network, the type of security, and the passcode for the hidden network.

The Wi-Fi symbol at the upper part of the screen indicates that the iPhone is linked to a Wi-Fi network. (To check, open Safari and go to a website.) When you return to the same place, your iPhone reconnects.

Connect to a Personal Hotspot

You can use the cellular internet connection of an iPad (Wi-Fi + Cellular) or another iPhone that is sharing a Personal Hotspot.

Navigate to Settings > Wi-Fi and choose the name of the device that will share the Personal Hotspot.

If your iPhone requests a passcode, input the one shown in Settings > click Cellular > Personal Hotspot on the gadget sharing the Personal Hotspot.

Connect your iPhone to a mobile network

If no Wi-Fi network is available, your iPhone automatically connects to your carrier's cellular data network. If your iPhone does not connect, do the following checks:

Check to ensure that your SIM card is active and unlocked.

Enter Settings > Cellular.

Ascertain that Cellular Data is enabled. On Dual SIM devices, hit Cellular Data and then confirm the chosen line. (There is a limit of one line for cellular data.)

How to Setup your iPhone 13

Whether you're new to the Apple ecosystem, or you're an addicted user of the product, you can set up your new iPhone easily. From the first "Hello" to the last step, here's all that you need to know about setting up your new iPhone 13 devices.

Chapter 4
HOW TO SET UP APPLE PAY

How to confirm identity for Apple Pay

Everything about Apple Cash Family (Setting up for kids, view transactions, etc.).

Set up Apple Pay Cash (How to send money, payments, add money to the card, request payment, etc.).

Set Up Apple Pay

Using Apple Pay is both faster and safer than using a physical card. Using the Wallet app to store your cards, you may use Apple Pay to make safe payments in stores, on public transportation, in applications, and on websites that accept Apple Pay. Take advantage of Apple Pay to make purchases from participating businesses while you're on the phone using Messages.

Assign your debit, credit, and prepaid cards to Wallet to get started with Apple Pay.

Add a Credit or Debit Card

In Wallet, press the Add Card button. You may be asked to check in with your Apple ID.

Choose one of the following:

Previous cards: Cards linked to your Apple ID, cards used with Apple Pay on other devices, and cards removed from your wallet are all acceptable options. Press the Continue button to continue, and then input the CVV code for each card.

Debit or credit card: You may either hold the iPhone in a position where your card is visible in the frame or manually enter the card info.

Transit card: Enter a location or card name, or scroll down to find transit cards in your region.

Set the Default Card and Reshuffle Your Cards

Your primary payment card is the first card you assign to Wallet. Push a card to the top of the stack to make it the default.

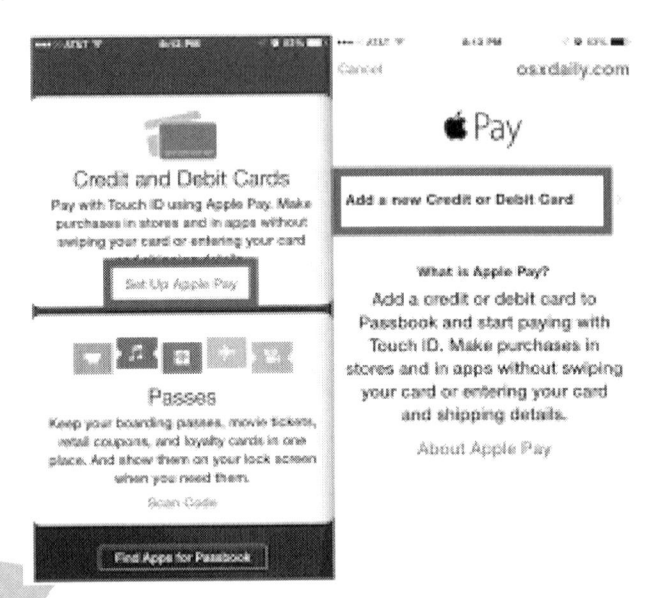

In Wallet, pick your default card, and hold down the card to bring it to the front of the stack.

To relocate another card, tap and hold it, then drag it to a new spot.

Utilize Apple Pay to Make Contactless Purchases

Apple Pay allows you to make safe, contactless payments at stores, restaurants, and more using your Apple Cash, credit, and debit cards saved in the Wallet app on your iPhone.

Pay Using Your Default Credit Card

Activate the side button by double-clicking it.

When your default card displays, either use Face ID or enter your passcode to authenticate with your iPhone.

The screen will show 'Done' and a checkmark when you bring the iPhone up to a few millimeters from the contactless reader.

Pay Using a Different Card Than the One You Normally Use

When your default card displays, tap it to access it and then select another.

Use Face ID, Touch ID, or your passcode to authenticate.

The screen will show 'Done' and a checkmark when you bring the iPhone up to a few millimeters from the contactless reader.

Utilize a Reward Card

When you use Apple Pay in participating stores, you may earn or redeem rewards.

Add your reward card to your wallet

Present your rewards card at the store's payment terminal by holding your iPhone near the contactless reader.

Apple Pay then uses your default payment card to complete the transaction. In select establishments, you may combine the use of your rewards card and payment card in a single transaction. In other establishments, you must wait until the terminal or cashier requests payment before proceeding.

To enable automatic selection of your rewards card when you enter a store, press the 'More' button on the card and then switch on Automatic Selection.

How to Make Apple Pay More Convenient When Shopping in Apps

When making purchases in apps, App Clips, and on the web in Safari, you may use Apple Pay if the button to make the transaction is available.

Pay Using Any App, an App Clip, or the Web

When payment, touch the Apple Pay button.

Examine the payment details.

You may update your credit card information, billing and shipping addresses, as well as your contact information.

Complete the transaction:

Double-click the side button, then either use Face ID or enter your passcode to authenticate with your iPhone.

Shop Using Your Mac's Safari Browser and Pay with Your iPhone

Before you begin, the following should be accomplished:

Enter your Apple ID credentials on both devices.

Each device should have Bluetooth turned on.

Ascertain that your iPhone is in close proximity and linked to a cellular or Wi-Fi network.

When you're ready to check out, complete these steps:

- On your Mac, begin the checkout process by selecting the Apple Pay payment method.
- Examine the payment details.
- You may update your credit card details, shipping address, and contact information at any time.
- Review the payment summary and finish the transaction on your iPhone.

If you do not wish to utilize Apple Pay on your iPhone to make Mac payments, navigate to Settings > Wallet and Apple Pay and disable Allow Payments on Mac.

Modify the Default Shipment and Contact Information for Your Account

To do so, navigate to Settings.

Go over to Wallet and Apple Pay.

Select from the following:

- Address for shipping
- Email
- Phone

Use Apple Cash

Your Apple Cash card in the Wallet app is replenished when you receive money over Messages. Using Apple Cash is as simple as using it anywhere you'd use Apple Pay. Additionally, you have the option of transferring your remaining Apple Cash to a bank account.

Set Up Apple Cash

Do any of the following:

Go over to Settings.

Click Wallet and Apple Pay, then turn on Apple Cash.

In Messages, send or receive a payment.

Use Apple Cash

You can use Apple Cash everywhere you use Apple Pay:

Apple Pay lets you send and receive money quickly and easily.

Make contactless payments with Apple Pay.

Pay in applications or on the web with Apple Pay.

Manage Your Apple Cash

In Wallet, tap the Apple Cash card.

View your current transactions, or scroll down to see all your transactions sorted by year.

Tap the More icon, then perform any of the following:

- Add money to your Wallet with a debit card.
- Transfer money to your bank.
- Update your bank account details.
- Request a statement.

Select whether to take all payments manually or automatically. Before the money is transferred to the sender, you have seven days to manually accept it.

View your recommended PIN. Face ID, Touch ID, or a secure password validates every payment with Apple Cash, so there's no need for a PIN. However, certain terminals may still ask you to input a four-digit number to complete the transaction.

IPHONE 13 USER GUIDE

Verify your identity for account service and raise your transaction limits.

Contact Apple Support.

Use Apple Card

In order to aid you in leading a more financially responsible lifestyle, Apple created the Apple Card credit card. Using the Wallet app on your iPhone, you can quickly sign up for an Apple Card and start using it with Apple Pay in stores, applications, and online throughout the world. Apple Card shows your recent transactions and balance in Wallet in a straightforward, real-time manner, and you can contact Apple Card support at any time by sending a text message from inside Messages.

Purchase an Apple Card

Press the Add Card option in Wallet, then tap Apply for Apple Card.

Submit your application by entering your information and agreeing to the terms and conditions.

Review the terms and conditions of your Apple Card offer, including the credit limit and annual %age rate, before accepting or declining the offer.

If you agree to the conditions, you may do the following actions:

Apple Card should be set as the default card for Apple Pay transactions.

Purchase a real Apple Card to use in locations that do not accept Apple Pay.

Utilize the Apple Card

Apple Card is compatible with all Apple Pay locations:

Utilize Apple Pay to make contactless purchases.

Apple Pay enables you to pay in applications and on the web.

Additionally, you may use Apple Card at establishments that do not support Apple Pay:

When using an app, browsing the web, or making a phone call: Tap the Card Details button to see the card number, expiry date, and password. Place an order using the details provided here.

Use the actual card at stores, restaurants, and other establishments.

Examine Transactions and Financial Statements

Tap Apple Card in Wallet.

Carry out one of the following:

Conduct an audit of your transactions: View your most recent transactions or scroll down to view all of your transactions by month and year.

Conduct a transaction search by tapping the Search button, entering the information you're searching for, and then tapping Search on the keyboard. Additionally, you may pick a recommended search, such as a category, merchant, or region, and then refine your search with extra text.

Take a look at weekly, monthly, or annual activity: To view your spending in categories such as Shopping, Food and Drinks, and Services, tap Activity (below Card Balance). To go to a different view, tap Week, Month, or Year. To view past periods, swipe right.

IPHONE 13 USER GUIDE

Obtain monthly financial statements: To view the amount, new spending, payments, and credits, tap Card Balance. To view your monthly statements, scroll down. To view the summary for a particular month, tap a statement. You may also download a PDF statement or export transactions to a CSV, OFX, QFX, or QBO file.

Pay Your Bills

Select the Payment option. Alternatively, hit the 'More' icon and then select one of the following:

Pay My Bill or Pay Different Amount: Select Pay My Bill or Pay Different Amount, enter the payment information (such as the account and date), and then verify using Face ID, Touch ID, or your passcode.

Pay a Bill: Drag the checkmark to change the payment amount or touch Show Keypad to input an amount, then press Pay Now or Pay Later, examine the payment information (such as the payment account), and finally verify with Face ID, Touch ID, or your password.

View Apple Card Data, Modify Settings, and Contact Support, Among Other Things

After tapping the 'More' button, perform one of the following actions:

Allow family members to use your Apple Card, establish transaction restrictions, and more.

Manage monthly Apple Card installments.

Conduct a review of planned payments.

View credit information.

Linked bank accounts can be added or removed.

Your actual card can be locked or unlocked, and you can request a replacement.

Edit the Express Transit configuration.

Modify your notification settings.

Your billing address should be updated.

Contact Apple Card Support through Messages, phone, or the Apple Card Support website.

Chapter 5

EXPLORE THE APP STORE

The App Store contains all Apps compatible with the device and makes it easy for you to find, buy and install them. The apps available for download in the store have been reviewed by Apple and deemed safe and suitable for use.

To locate apps, click on the App Store and type in word(s) relating to the app in the top-left-hand corner of the Apps store window and then press 'Enter.' A list of apps will be populated on the screen, and you can then click on the apps' name or icon to see the description, supported languages, file size, previews, ratings, compatibility with Apple devices, and to view existing user ratings and reviews.

Download and purchase apps: To download a particular app, tap on the button that displays the price or 'Get' and then click on the button one more time to buy the app or install (if the app is free). To purchase an app or in-app content, you will need to enter your Apple ID or use your Face ID. Sometimes, the app you wish to download has previously been purchased by other family members if you are part of a Family Sharing group. While in the App Store, click on your name on the bottom-left-hand side of the page, and a list of all apps purchased using your Apple ID will be displayed. Next, to download the App, click on the 'Purchased by' > family member's name > iCloud status icon beside the app.

To share an app, click on the app to see its details and then a sharing option.

Hide a purchased app: To hide a purchased app, hold the pointer over an app, click the 'Show more' button and then choose Hide Purchase. To unhide a hidden apps click on View Information > Manage > Unhide > Done.

Update installed apps: The moment an update is available for any of your downloaded Apps, a notification will be displayed in the Notifications Center, and a badge will be visible on the App Store icon in the Home screen (with a number displayed indicating the number of updates available). From the updates pane in the Apps store, you have the option of updating all Apps (click Update All) or updating individual apps (click Update beside each app you intend to update). You can also update an app once you receive a notification that an updated version of the apps is available or from the Apps store > Updates.

If you prefer that all your apps, be automatically updated, App store > Preferences > Automatic Updates.

Re-download an uninstalled app: To reinstall a previously installed app that you uninstalled or deleted, navigate to Apps store > re-download the app by clicking the re-download button.

Uninstall an app: To uninstall an app, drag the app from the Applications folder to the Trash. Once the Trash is emptied, the app is permanently removed. However, you can still get it back from the Trash before emptying by selecting Trash > File > Put Back. Also, from the Launchpad, hold an app till the other apps begin to move gently side to side and click the visible Delete button (for apps purchased from App Store).

You can subscribe to Apple Arcade in the App store to have access to a huge collection of iPhone games. App store > Arcade > Start a free one-month subscription or Start a monthly subscription > Subscribe > Authorize with Apple / Face / Touch ID.

Note that the availability of an app in the App store is dependent on the country or region you are located in.

Lots of games to play with your friends on the iPhone

Go to App store > Games or Arcade or Enter the name of the game in the search bar and download it as you would other apps. You can access your downloaded games and progress on other devices you are signed in to with the same Apple ID/

Books

Book is an app that contains books (both visual and audio) that enables you to read, listen to, arrange existing books and at the same time buy new books.

Buy books: Go to the Books > Book store and type the name of a book you wish to buy. Select a book, click on the book's price and then buy a book / get book (if the book is free) or pre-order. To keep a book for later, click the Options arrow and then Add to Wish List.

Find an audiobook: Go to Books > Audiobook store, after which you can buy the book gift the book after clicking the options arrow.

Read a book: Navigate to Books > library and then select the book by double-clicking the book to open.

Have a book read to you: While you are doing something else, you can have a book read to you. Go to Edit > Speech > Start Speaking. If the book comes with a Read Aloud feature, simply click the Speak button at the top of the book (in the toolbar).

Improve the reading conditions at Night

To improve reading conditions, choose view > Theme and Night or Appearance > click the Black Circle.

Bookmark a page: When reading a book, click the Bookmark button to ensure the page is saved and you do not lose track of the page you stopped.

Restrict the access to certain Books

To limit access to a book, go to Settings > Screen Time > Content and Privacy and then turn on Restrictions. Two actions can be done here; either to Disable Book Store and AudioBook store (Apps > Deselect Books Store) or Restrict Books with explicit content (Stores > Deselect Explicit Books)

To delete a book: Once a book is no longer required, delete the book by Book > Library, select the book and press the 'Delete' key.

iTunes

iTunes store allows you to access the world's largest music, films, and TV shows collection, listen to your collection of songs and organize the songs and albums you buy on iTunes in your personal library.

Once you buy or download music, it is updated in your music library, and any song that is not already in the Store is added to iCloud.

To add or download music to your device, launch the iTunes app, search the catalog (input a keyword that relates to the song in the search field). When you see the song of your choice in the search results, select the 'More' button next to it and add it to the library or Download. Downloading a song allows you have access to it even when you are offline (not connected to the internet). You can also view recommendations of songs that were curated for you based on your listening history and preferences and 'What's new' to find new songs.

Listen to Songs: To play songs, launch the music app, navigate to the song you wish to listen to, click on it and Play or Click on the song. You can use the controls to pause, repeat, play songs in a defined order or shuffle songs.

Share music library with others: You can share with other users that you authorize to (with a password).

iTunes allows you to buy, rent and watch all your favorite movies and TV shows from any of your devices.

Search for Content: You can browse for content by clicking the Movies, TV shows, or Kids tab in the menu and then pick the genre and look through. When you find the content, you wish to watch, you have the option of buying or renting. You can also share with up to six family members through Family sharing.

How to Automate and Schedule Focus Modes

You can engage a Focus mode on your iPhone whenever you want to block out distractions and focus your attention on something specific, but you can also plan Focus modes to activate at certain periods of the day, such as during work hours or as you prepare to sleep. Furthermore, you can configure a Focus to come on automatically when you enter a certain place, such as your office, or when you open a specific app, such as Books or the TV app.

Open Settings and then hit Focus.

Select the Focus mode for scheduling.

Tap Add Schedule or Automation under "Turn on Automatically."

Depending on your use case, choose Time, Location, or App.

If you choose Time, specify the period using the From and To selectors, as well as the days on which the Focus should appear. If you chose Location, use the input box to search for or enter an address, then touch Done. If you choose App, touch the applications in the list that you want to activate the Focus on, and then hit Done.

That is all. Once you've scheduled a Focus or created automation, you'll find it listed under "Turn On Automatically" in Settings -> Focus, where you can enable/disable it or totally remove it.

Create a Focus

With Focus, you can configure your device to assist you in being present in the moment by establishing a personalized Focus or choosing one depending on the situation, such as during work hours or as you prepare for bed. Furthermore, it is possible to build Home Screen pages with applications and widgets that are relevant to certain times of concentration in order to show just relevant apps and avoid temptation.

Begin by launching the Control Center and selecting the Focus button.

At the lower part, touch the plus sign-indicated New Focus button.

To create a new Focus, click Custom.

Give your customized Focus a name and choose a color/emoji/icon to distinguish it, then press Next.

On the next screen, touch Add Person to choose the individuals from whom you want to receive alerts while Focus mode is activated. Also, you can choose whether to accept calls from No One, Everyone, Favorites, or Entire Contacts, each of which is a separate call-specific choice.

Allow [X] Individual or Allow None.

On the next page, touch Add App to choose the applications from which you want to get alerts while Focus mode is activated.

Allow [X] Applications or Allow None.

Allow Time-Sensitive Notifications when the Focus is activated or select Not Now to decide later.

Tap Done to complete the process of generating the custom Focus.

Switch on Focus

Go to the Control center.

Click the focus icon, then choose the focus.

Press the 3-dot icon to specify focus duration.

Delete Focus

Open Settings.

Choose the Focus that you want to remove.

At the lower part, hit Delete Focus.

Customize Focus

If you wish to modify an already-created Focus, go to Settings -> Focus and choose the desired Focus mode. There are many settings accessible under the Focus choices. By clicking the Focus's name and icon at the top of the menu, you can modify its name and icon, as well as the people and applications from whom you want to get alerts.

Below that, an option for enabling/disabling Time-Sensitive Notifications is visible. When this option is enabled, individuals and applications will be able to alert you quickly, even if the Focus is enabled. Under "Options," you'll find Focus Status - enabling this option informs individuals who contact you that you've disabled notifications. Furthermore, you'll notice an option to Hide Notification Badges on app icons and a toggle to Hide Specific Home Screen Pages. Additionally, you can opt to dim the Lock Screen's look and to show any silent alerts you get on the Lock Screen.

You'll see a Smart Activation option under the last column, "Turn on Automatically." By enabling this, the chosen Focus will be activated automatically at appropriate moments during the day, depending on signals like your location, app activity,

IPHONE 13 USER GUIDE

and more. If this is too ambiguous, you can use the alternative Add Schedule or Automation option to have the Focus switch on automatically at a certain time, place, or when using a specific app.

Enable Time-Based Notifications

Time-sensitive notifications are a new notification type in iOS 15, and they can be permitted to escape Focus mode. Notifications classified as time-sensitive are those that are urgent, such as a knock at the door, a meal delivery, or a ride pickup.

On your iPhone, open the Settings app and then hit Focus.

Select the Focus mode you want to permit time-sensitive alerts.

To turn on Time-Sensitive Notifications, toggle the switch labeled Time-Sensitive Notifications to the green ON position.

Activate Smart Activation

For instance, if you create a custom Focus mode named "Studying" and spend your weekdays studying using the Books app, at the library, or right after lunch, Smart Activation can learn your routine and intelligently activate the Focus mode you created without your intervention. Configuring Smart Activation for a particular Focus mode is straightforward once you understand how. Consider the steps below.

On your iPhone, open the Settings app and then hit Focus.

Choose the Focus mode you'd want to automate.

Select Smart Activation from the "Turn On Automatically" menu.

To turn on Smart Activation, toggle the switch next to it to the green ON position.

The Settings App

Now, let's talk about the Settings app. The Settings application, which has a gear-shaped icon, is the app responsible for controlling everything. With the Settings app, you can configure your smartphone to function in a certain way you want to, grant permission for data sharing with apps, personalize your smartphone to suit your taste, and determine how most applications should function.

The settings app also helps you protect your data by controlling your privacy and security. If you can't find the Settings app on the Home Page, you can tap anywhere and swipe downwards. The swipe gesture will reveal a search bar that Apple calls Spotlight Search, where you can search for any application on your device by either spelling it out or tapping on the microphone icon to the right of the search bar and using the Voice Search feature. You can then type in "Settings" into Spotlight Search, and the application will come up.

When in the Settings app, your name should appear just below the "Settings" text, except you have yet to log into your iCloud account or register one. To log into your iCloud account, tap on the account icon and log in. It's not a necessity, but not having an iCloud account prevents you from accessing some features of your iPhone, like iMessage, FaceTime, or having access to the App Store (which is necessary to download apps).

On the Settings app screen, you can modify the settings configuration of various features like Bluetooth, Wi-Fi, Notifications, Control Center, Display and Brightness, Home screen, etc.

Notifications

If you want apps that you have installed to send you notifications, you can go ahead and monitor and kind of change your notifications settings. So, for instance, if you don't want the App Store to send you notifications, you can turn off the toggle switch for Allow Notifications for the App Store.

You can go down the list and adjust the settings of all the different listed features to your preference. You can change your vibrations, sound settings, ringer and alerts, ringtone, keyboard clicks, haptics, etc.

Do Not Disturb

The Do Not Disturb feature helps to mute the sounds made by different apps to prevent them from interfering with work, school, meetings, or quiet time.

Screen Time

Screen Time is a new app that allows you to view which apps you're using the most and how much time you have spent actively using that application on your smartphone. So, Screen Time is a record of how much time you spend actively using your iPhone.

Storage

Right here is another big one: iPhone storage. With this, you can monitor how these apps make use of your storage. So, if you don't like how certain apps are using your storage, you can always modify them accordingly.

General

The General setting in the Settings app is one of the most popular destinations for people who access the Settings app. If you tap on it, you can monitor and modify lots of things.

When you tap on General, the first feature you will find is the About feature.

In the General settings screen, there is also a feature called Background App Refresh, which helps you disable an app from self-refreshing in the background. There are toggle switches against each installed application, and you can turn off the toggle switches for the applications you don't want to self-refresh.

Turning off the Background App Refresh for these apps will help you conserve some battery power, as the constant refreshing makes the app use more than its fair share of battery power.

Reset

There is also a Reset feature in the General settings screen, and its purpose is to return your device to factory conditions. If you ever need to reset your device (perhaps you are done using the device, or you want to trade it up for a newer model), you can reset the device and wipe off all the stored data. You can also use the Reset feature to reset the entire settings app to factory conditions using the Reset All Settings option. This chosen option doesn't wipe off all the stored data; all it does is return all your settings to their default.

The Control Center

The Control Center settings allow you to change the controls in the Control Center. You can change the arrangement and also add and remove some specific controls.

IPHONE 13 USER GUIDE

Features under "Display and Brightness" settings

Dark Mode

If you go back to the Settings screen and tap on Display and Brightness, you can now turn on the Dark Mode feature. To turn on dark mode, you can tap on the dark mode icon at the top of the Display and Brightness screen, and it will make all the pixels black, which makes videos more pronounced. You can also set it up automatically to automatically switch to light or dark mode at different times of the day.

True Tone

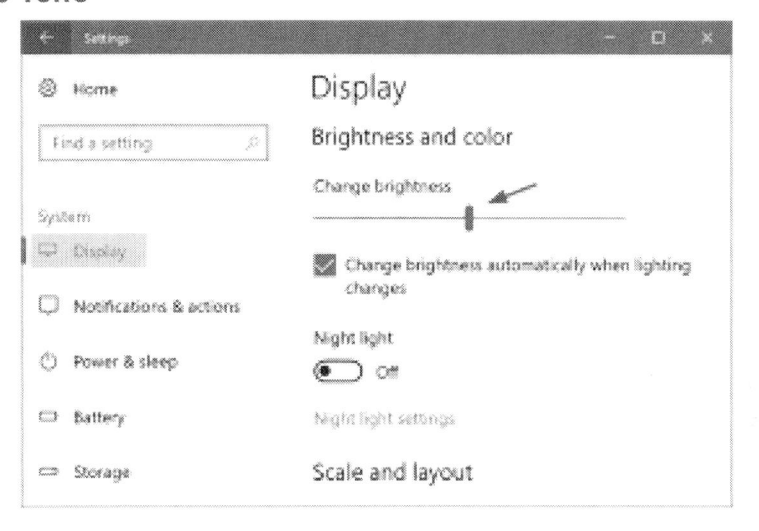

True Tone is another helpful feature in the Display and Brightness settings screen. True Tone self-adjusts the display light to the ambient lighting condition of its immediate environment, thus giving you color that is as natural as possible.

Night Shift

There is also a toggle switch for Night Shift, a feature in iOS 15 that automatically changes the various colors of your display to make them softer and warmer to protect your eyes at night. You can turn on this toggle switch in the Display and Brightness settings screen to enable the feature.

Auto-lock

Auto-lock is pretty much how long you want your display to remain on before it self-locks. There are different durations in the auto-lock feature, ranging from "30 Seconds" to "Never." So, if you would like your display to self-lock after 30 seconds of inactivity, you can select that option, and your phone will auto-lock after every 30 seconds of dormancy.

If you go down the Settings screen, besides the preloaded settings, you will find applications you've downloaded to your device, and you can tap on any of them to alter their settings.

The Home screen

Now, let's say you downloaded an app to your Home screen that you want to take off the Home screen. To do this, tap and hold any empty part of the Home screen, and this will launch the device into jiggle mode, which is the mode where all apps become wobbly and editable. You can then touch and hold the app you do not want on the Home screen, delete it, move it to another Home screen, or hide it in the App Library.

Jiggle Mode

Jiggle Mode is an iOS feature where all Home screen applications and widgets become editable when you tap and hold an empty portion of the Home screen for about two seconds.

To exit the jiggle mode, tap on "Done," located in the right corner of the display beside the notch. The Home screen will return to normal.

IPHONE 13 USER GUIDE

Now to get back into jiggle mode, go on and hold down on the long-press and any empty part of the Home screen. Now, tap on the button on the top left corner of every app icon when in jiggle mode. Three options will appear when you tap on it: Delete App, Move to App Library, and Cancel. Delete and Move Apps

If you want to delete an app, then click on the "Delete App" option. Selecting this option will remove the application entirely from the device. If you want to move the app to the App Library, so it's hidden from the Home screen but remains in the App Library on your device, then tap on that option. The "Cancel" button is to exit the menu should you change your mind.

If you're in any app and want to return to the Home screen, swipe upwards from the base of the screen. The upward swipe will take you back to the Home screen, but it doesn't close the application you were using previously.

Spotlight Search and Today View

Spotlight Search, a search bar at the top of the iPhone's "Today View" screen (the last screen to the left of the Home screen), helps you quickly find any information you're looking for on your iPhone. Once you open the Today View screen and swipe downwards, the Spotlight Search bar comes up. You can then search for an application, a media file, or any other content on your device.

The App Store

On the App Store, it's straightforward to download and install apps. To download apps on the app store, tap on the search icon at the bottom right corner of the display in the App Store. A search bar will come up, and you can search for whatever app you seek.

You can also update existing apps through the App Store.

Apps that you download automatically install by themselves after the downloading is complete. At the end of the download and install process, you will see a virtual Open button pop up, and you can launch the app directly from the App Store.

Chapter 6

IPHONE 13

Here's how you can force close open applications on your iPhone 13. If, for instance, an app is acting strange or consuming your battery power irrationally, or perhaps you don't want the app to be refreshing itself or covertly performing tasks, you can force-close the application. If you've tried force-stopping a running application on the iPhone 13, you may have observed that the standard swipe-up gesture for shutting down applications doesn't work; it instead returns your device to the Home screen.

The iPhone 13 (powered by iOS 15) offers a new way of exiting apps that involves a two-step process comprising a gesture and then long-pressing. It will take some getting accustomed to, but the eventual result will be the same: you will be able to force stop iOS 15 applications running.

How to Force Close Active Applications on the iPhone 13 from the App Switcher

On your iPhone 13 device, swipe up from the bottom of the screen and then pause when the app cards appear in the center of the screen.

Then, long-press any app preview card until the force-stop icon (symbolized by ⊖) appears in the top-left corner of each app card, as shown below:

To force close the application, tap ⊖.

If you wish to force close other open applications also, tap the red minus (-) sign in the top-left corner of the respective apps to force close them.

To leave the multitasking mode (app switcher), swipe up from the bottom of the display.

To force close apps on iPhone 13, scroll up to open the app switcher, long-press any apps to bring up the force-stop icon ⊖, then tap on it to force-stop the application. Alternatively, once the minus signs on the app cards come up, swipe up on any app you want to shut down. The press and hold gesture is similar to how you uninstall apps from your iPhone's Home screen, so if you've been using iOS for a while, you should be familiar with it.

It is important to note that forcing an application to close is not the same as uninstalling the application.

How to create a folder on the Home Screen

Once the iPhone boots, the landing page is the Home Screen, where basic functions such as opening apps, searching for and organizing files and documents reside. Basically, once you are at the Home Screen, you can swipe/navigate to find any files, app, or folder and access any functionality of your choice.

You can also pin essential widgets to the Home screen to ease working with multiple apps at the same time. Go to the Home screen and swipe upwards from the lower part of the screen for iPhones with Face ID, or press the Home button for iPhones with Home button.

Apps are arranged by categories, and the frequently used apps are located at the top of the screen for easy location and usage.

Widgets on the Home Screen

Widgets for Today View displays current information from your apps at a simple glance. Information such as today's headlines, weather, calendar events, and much more.

To move a widget from today's view to the Home screen, open the Today view and find the widget you want to move. Press and hold the widget until its icon starts to move from side to side gently. Next, move it off the right side of the screen and place it on the Home screen and click Done.

To add a widget to the Home screen, navigate to the Home screen page where you want to add the widget. Then, press and hold the Home screen background until the apps begin to jiggle. Click on Add + at the top of the screen to open up the widget gallery. Find the widget you intend to add to the Home screen, click on it, and swipe to the left for the various size options.

To remove a widget, touch and hold the widget for the quick

IPHONE 13 USER GUIDE

actions options to be displayed. Then, click on the Remove widget and then click Remove.

Wake and unlock your iPhone

To conserve battery power, the iPhone turns off the screen display, locks the screen, and sleeps when not in use. At any time, you can wake and unlock the device.

To wake the iPhone from sleep, you can either press the side or Sleep/Wake button, raise the iPhone or tap on the screen.

To unlock the iPhone with Face ID, tap the screen display or raise the iPhone up to wake it from sleep and gaze at the screen. The iPhone will be unlocked provided you have set up Face ID earlier.

To unlock the iPhone with Touch ID, tap the Home button with the finger you have configured earlier for Touch ID.

When the iPhone screen is locked, the time, date, and recent notifications appear when you wake the device. Also, you can launch the Control center and camera and notification previews even when the screen is locked.

Restart or Turn off the iPhone

You may restart the iPhone when it is not working as expected. When restarted, the device switches off and comes on again. Restart your iPhone by pressing and releasing the volume up button and afterward the volume down button. Press and hold the side button and then release the button when the Apple logo appears on the screen.

To turn off the iPhone with Face ID, you can tap and hold the side button together with the volume button until the sliders become visible. At this point, pull the top slider. With an iPhone with a Home button, tap and hold the side button or the Sleep/Wake button and then pull the slider.

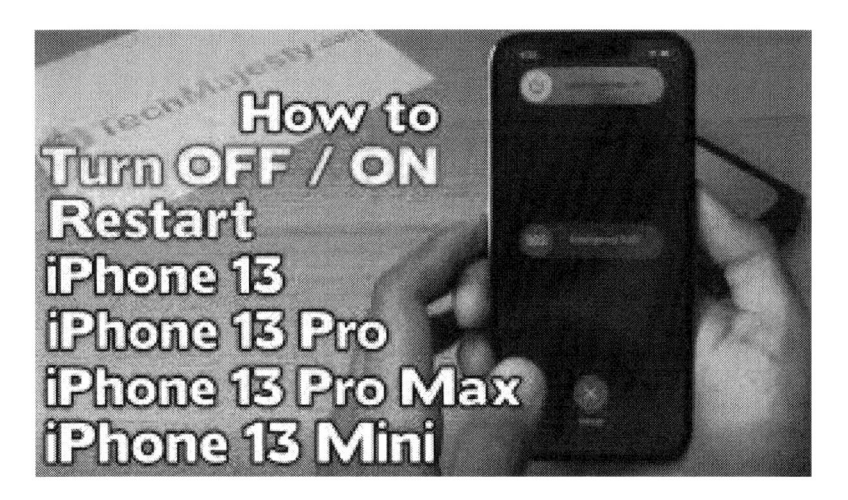

IPHONE 13 USER GUIDE

Chapter 7

APPLE ID AND ICLOUD

Apple ID is an ID that you can use across a range of Apple devices to access Apple services and applications, while iCloud is a secure repository to store your photos, videos, documents, music, apps, files, books, and much more across a range of multiple Apple devices. iCloud allows you to keep and share photos, reminders, calendars, and other things easily with friends and family. Even in the event of a loss, you would not lose your files as long as you can access iCloud on another device. iCloud gives you a free email account and about 5 GB of free storage for your files. However, purchased content would not deplete the allocation of 5 GB.

For more space, you can upgrade your iCloud storage on your iPhone. Settings > your name > iCloud > Manage Storage > Change Storage Plan. Here, you can also switch on the features that you intend to use, such as Messages, Reminders, Safari, and so on.

To sign in with your Apple ID, Settings > Sign in to your iPhone > Input Apple ID and password.

Synchronize your iPhone with your computer to keep the contents of the iCloud up to date between the two devices. Contents such as songs, movies, podcasts, photos, videos, contacts, and calendars.

IPHONE 13 USER GUIDE

To sync your iPhone and Mac, connect the two devices with a USB cable. Locate the Finder sidebar on your Mac and click on your iPhone's icon. Select the content you want to sync on the top of the window and click on 'sync' onto 'device name.' Sync all content types as you desire and click 'Apply'. Subsequently, whenever you connect the two devices, the contents are synced.

You can also enable Wi-Fi syncing. First, connect your iPhone and computer using USB. Locate the Finder sidebar on your Mac and click on your iPhone's icon

Transfer Data from an Android Device to your new iPhone

While setting up your iPhone for the first time, you can also migrate data from an Android device. Make use of the 'Move to iOS app' when you are setting up for the first time. Subsequently, you would need to clear all the data on the iPhone and start over or migrate the data manually. The steps to follow are outlined below:

Download the 'Move to iOS' app on the Android device with Android version 4.0 or later. Also, switch on Wi-Fi.

At the same time, on the iPhone, navigate to the Apps and Data screen and then click 'Move Data from Android.'

Launch the Move to iOS app and follow the prompts on the Android device.

Setting the Volume on the iPhone: when listening to content with audio output on the iPhone, there are two options to adjust the volume. You can use the volume buttons at the side of the iPhone to adjust the volume of audio outputs or Ask Siri to process the Command. You can also manage the volume in the Control Center.

Everything about Widgets on Home Screen

Apple's introduction of widget support remains one of its most significant changes to iOS 15. You can get simplified, timely updates from your frequently-used applications with widgets. You can add widgets to your iPhone 13's home screen (since it runs the iOS 15 operating system) to stay abreast of all the new and essential information that you care about knowing.

A widget is a self-contained mini-program that runs on an iPhone's Home screen and performs basic tasks like displaying weather data or news highlights. Widgets are dynamic and self-update, saving you the time and effort of doing so manually.

Several widgets are now available on the App Store. One example is the Weather widget (which displays the current weather in a chosen area), the Battery widget (which tracks the battery percentage of all Apple devices connected to your iPhone in real-time), and others.

Different applications have different widgets, and you must prioritize which ones will provide you with the best value to maximize your limited Home screen real estate.

There are three widget sizes to choose from:

- A large square-shaped widget
- A medium rectangular-shaped widget
- A small square-sized widget

In iOS 15, you may create a scrollable stack by stacking widgets of the same size on top of each other.

How to add widgets to your iPhone's Home screen:

Long press a widget or a blank part on the Home screen until your device switches to jiggle mode (where all the applications begin to vibrate). In the top-left part of the display, tap the + icon.

Choose any widget you like from the suggested widgets, then choose one of the three available widget sizes (small, medium, and large).

Then click "Add Widget."

To implement your changes, tap on "Done."

Stack your widgets!

You can use the Home screen and Today View screen more efficiently on your iPhone 13 series device by bundling your widgets into stacks. You can use Smart Stacks (which auto-scrolls the stacked widgets for you), or you can design a personalized widget stack.

Building a Smart Stack

Smart Stacks are a pre-configured arrangement of widgets that presents the appropriate widget as dictated by influences like where you're located at the time, the task you're engaged in, or the period of the day. All through the day, your Smart Stack self-changes the widgets it displays to offer you the most period-appropriate information. To build a Smart Stack, follow these steps:

Launch the Home screen into jiggle mode by tapping and holding on to the blank part of the Home screen.

In your display's top-left corner, tap the Browse to the bottom of the page and choose Smart Stack.

Then select Add Widget from the options.

Make your personalized stack of widgets.

On your iPhone's Home screen, press and hold an application or a blank region until everything jiggles.

Drag and drop one widget on top of another. You can pile as many as ten widgets over each other.

Press the Done button when you are through with your stacking.

Modify a stack of widgets

Choose Edit Stack from the drop-down menu. With this, you can then reorganize the widget arrangement in the stack. Simply drag the sign to do this. If there are specific widgets you want to display during the day, you can enable Smart Rotate. Alternatively, swipe leftward over a widget to remove it.

When satisfied, tap the ✕ icon.

Get rid of widgets

If you need to remove any widget, follow these steps:

Long-press the specific widget you wish to remove.

Select "Remove Widget" from the drop-down menu.

To affirm, select "Remove" once more.

How to Take a Screenshot

Among the first things you should learn when using your iPhone 13 Pro Max is how to take a screenshot. Fortunately, the process is relatively straightforward, and understanding it takes just a minute!

To take a screenshot, click the volume up button and the power button (or side button) simultaneously on the screen you want to capture.

IPHONE 13 USER GUIDE

For every screenshot you take, you should see the captured image miniaturized at the bottom of the display.

A snapshot immediately appears in the bottom left corner of the active display; it disappears after a while if no action is taken.

Tap the snapshot to make changes, or tap-and-hold to share the screenshot right away with an application via AirDrop or other sharing methods.

You'll see the screen below if you wish to edit or modify the screenshot. You'll be provided with the option to either "Save to Photos" or "Delete" the screenshot if you tap the "Done" button in the top-right corner.

Chapter 8

EVERYTHING ABOUT FACETIME CALL

Using Facetime on the iPhone, Take a Live Photo

A FaceTime Live Photo captures an important part of your conversation while you're on a video chat with someone (not available in all countries or regions). The camera records everything that happens shortly before and after you take the picture, including the audio, so that you may see and hear everything precisely as it happened later on in the video or audio.

IPHONE 13 USER GUIDE

You can take a FaceTime Live Photo by first enabling FaceTime Live Photos in Settings > FaceTime, and then either:

When in a conference call with another party, tap the Take Picture button.

On a FaceTime call with a group of people: Hit the tile of the subject you wish to capture, then tap the Full-Screen button.

Use Other Applications During a Facetime Conversation

While on a FaceTime conversation, you may use other applications to accomplish tasks such as searching for information or performing a computation.

To open an app, navigate to the Home Screen and press its icon.

Tap the green bar at the top of the screen (or the FaceTime symbol) to return to the FaceTime screen.

Begin a Facetime Call with a Group of People

Near the top of the screen on FaceTime, hit New FaceTime.

In the top entry area, enter the names or phone numbers of the individuals you wish to call.

Additionally, you may press the Add Contact icon to launch the Contacts app and add contacts directly from there. Alternatively, you may access suggested contacts by tapping on their names in your call history.

To initiate a visual call, press the FaceTime button, and to initiate an audio call, tap the Call button.

On the screen, each participant is represented by a tile. When a person talks (verbally or through sign language) or taps a tile, it becomes more visible. Tiles that are too large to fit on the screen are shown at the bottom in a row. Swipe through the row to discover a participant you missed. (If no image is available, the participant's initials will display on the tile.)

Creates a Blurred Background in Portrait Mode

Like the Camera app's Portrait mode, you can turn on Portrait mode to make the background disappear and draw the viewer's attention to yourself.

Tap your tile while you are on a FaceTime chat.

In your tile, select the Blur Background option.

Again, touch the button to exit Portrait mode.

Toggle the Back Camera On

While on a FaceTime call, press your tile and then the Back Camera button.

Tap the Flip to Back Camera button once more to revert to the front camera.

When utilizing the back camera, you may magnify the image by touching 1x. Re-tapping restores the picture to its default size.

In Facetime on the iPhone, Block Undesirable Calls

You can disable FaceTime calls from unknown callers inside the FaceTime app.

Tap the Info button next to the contact's name, phone number, or email address in your FaceTime call history to view their information.

Scroll to the bottom and hit Block This Caller, followed by Block Contact.

To block a contact, choose it.

Delete a Call from the History of Your Calls

Swipe left on a call in your call history in FaceTime, and then hit Delete.

How to Drag and Drop screenshot

Make a screenshot as normal by tapping the Side and Volume Up buttons simultaneously.

Hold down the screenshot thumbnail in the bottom-left corner of the screen for a few seconds until the white frame around it disappears.

Tap the app in which you wish to use the screenshot with another finger. In this example, we're going to launch the Photos app, but you could also open Files, Messages, Mail, Notes, or something else.

Navigate to the location where you wish to utilize the screenshot.

Move the screenshot to the desired location, then release your finger to drop it in place.

How to Activate a Focus

It is simple to turn a Focus on. Simply open the Control Center, press the Focus button, and then choose the Focus you wish to enable. You may also activate it by tapping the ellipsis (three dots) button. For one hour, till tonight, or until I leave this place.

How to Focus

Tap the Focus button to open the Control Center.

Tap the New Focus button (represented by a plus symbol) at

IPHONE 13 USER GUIDE

the bottom.

To make a new Focus, choose Custom.

Give your customized Focus a name and select a color/emoji/icon to make it stand out, then press Next.

On the following page, tap Add Person to choose the individuals from whom you wish to receive alerts while the Focus mode is active. You may also select to receive calls from Everyone, No One, Favorites, or All Contacts as a separate call-specific option.

Allow [X] Person or Allow None to proceed.

Tap Add App on the following page to select which applications you want to get alerts from when the Focus mode is active.

Allow [X] Apps or Allow None to proceed.

On the following page, select Allow Time-Sensitive Notifications while the Focus is activated or tap Not Now to make a later decision.

To complete the custom Focus, tap Done.

How to Personalize a Focus

If you wish to customize a previously configured Focus, click to Settings -> Focus and choose the Focus mode in question.

There are numerous settings accessible in the Focus choices. You may modify the name and icon of the Focus by pressing it at the top of the menu, as well as the people and applications from whom you want to get notifications.

Below is a toggle for enabling/disabling Time-Sensitive Notifications. People and applications will be able to alert you instantly if you choose this option, even if you have the Focus switched on.

Under "Options," you'll see Focus Status enabling this allows applications to notify those who contact you that your notifications have been hushed. You'll also find the option to Conceal Notification Badges on app icons, as well as the Custom Page toggle, which allows you to hide specific Home Screen pages. You may also opt to dim the Lock Screen's look and show any silent alerts you get on the Lock Screen.

You'll see a Smart Activation option in the last section, "Turn on Automatically." When you enable this, the selected Focus will be activated automatically at relevant moments throughout the day, based on signals such as your location, app use, and more.

If this is too ambiguous, you may utilize the Add Schedule or Automation option to have the Focus switch on automatically at a specific time, location, or while using a specific app.

How to Automate and Schedule Focus Modes

Open the Settings app on your iPhone, then choose Focus.

Choose the Focus mode you wish to plan.

Tap Add Schedule or Automation under "Turn On Automatically."

Depending on your use case, choose Time, Location, or App.

If you choose Time, enter the duration using the From and To selections, as well as the days on which you want the Focus to appear. If you choose Location, use the input area to search for or enter an address, then press Done. If you choose App, press the applications in the list for which you want the Focus to be activated, then tap Done.

IPHONE 13 USER GUIDE

That's the only thing there is to it. Once you've scheduled a Focus or created automation, it will appear in Settings -> Focus under "Turn On Automatically," where you may enable/disable it or delete it entirely.

How to Allow Time-Sensitive Notifications to Override Focus Mode

By following the instructions below, you may determine whether time-sensitive alerts can be received during particular Focus modes.

Open the Settings app on your iPhone, then choose Focus.

Choose the Focus mode to which you wish to enable time-sensitive alerts to pass.

Toggle the Time-Sensitive Notifications switch to the green ON position.

How to Get Augmented Reality Walking Directions in Maps

In a nod to Google Maps, the new AR mode may use your iPhone's back camera to overlay walking directions onto the actual environment, making it simpler to see where you need to go in densely populated places and eliminating the need to look down at your smartphone as you go.

Begin by starting a walking path, then raise your iPhone and scan the structures around you when asked. The step-by-step directions will show automatically in the AR mode, which should make it easier to go where you need to go, especially in difficult-to-follow circumstances.

The AR functionality will be accessible in key supported cities such as London, Los Angeles, New York, Philadelphia, San Diego, San Francisco, and Washington DC beginning in late 2021.

To receive augmented reality walking instructions, your iPhone must have an A12 processor or later. The A12 chip was initially utilized in the iPhone XS, XS Max, and XR, which were introduced in 2018; therefore, iPhones manufactured after 2018 are compatible with AR functionality.

How to Use Siri to Share Whatever Is on Your Screen

One benefit of Siri's improved contextual awareness is its ability to let you share whatever is on your iPhone screen with someone else via Message, whether it's a webpage in Safari, a song in Apple Music, a photo, or even the local weather forecast.

Say "Hey Siri," then "Share this with [person]" at any moment to share something. Siri will take action and confirm your request by asking, "Are you ready to send it?" At that point, you may either answer yes/no or add a remark to the message using the input area before clicking Send.

If the information cannot be transmitted directly, such as the weather forecast, Siri will snap a screenshot and email it instead. Simply say, "Share this with [person]," and Siri will capture the screenshot and confirm the request with you in the same manner.

How to Use Apple Maps to Locate Transit Stations Near You

How to Use the Dedicated Apple Maps Guide Section to Find Fun Things to Do

Furthermore, Apple has included a dedicated Guidelines Home that includes editorially chosen guides with tips on things to do in a place where you reside or travel.

Drag your finger up the main menu card in the Maps app to browse Apple's instructions at any moment. Scroll horizontally in the section under "Editor's Picks" to see which guides Apple is presently emphasizing. Alternatively, you may travel to the main Guides Home by tapping Explore Guides.

The Guide's Home is structured vertically, with Editor's Picks at the top, seasonal guides, the most recent guides, and cities following. If you keep scrolling, you'll come across a section where you may browse by publisher. By tapping the chevron at the top of Guides Home, you may choose between city guides for the entire globe and city guides for specific continents.

When you open a guide, you'll see a map that highlights all of the locations mentioned in the editorial below. Tapping on a location provides further information such as opening hours, reviews, directions, and so on, or you may scroll down the guide and read the short descriptions of each location.

You'll notice choices at the top of the editorial section to go to the publisher's website, share the guide, or save it. If you save it, it will be added to the "My Guides" area of the main Maps menu card.

Use your iPhone or iPad to launch FaceTime.

Press New FaceTime and enter the names of the people with whom you wish to share your screen, then tap the FaceTime button. Alternatively, you may begin a video conference by selecting a recent contact.

When the call is connected, touch the SharePlay button in the new control panel in the top-right corner of the screen.

In the dropdown menu, choose 'Share My Screen'. Screen sharing should begin after a three-second countdown.

After FaceTime screen sharing has begun, you may go to any app you want to share with the callers. To show that FaceTime screen sharing is active, a sharing symbol will stay in the top-left corner of the screen, and you may press it to open the FaceTime control panel.

You may quickly slide away from the active caller's face for extra screen space, and then swipe them back into view. If you're watching someone else's shared screen, you'll see their name immediately below the top-left icon, as well as options to write them a message, love what they're sharing, and share it with someone else.

You may also use the SharePlay interface to listen to music or view movies and TV together. You may watch movies or TV episodes while on the phone, and everyone on the call will see the same synchronized playback and controls.

Unlocking with an Apple Watch Isn't Working? Here's How to Solve the Issue

When one glance at iPhone locked and the Face ID detects that mask is worn, then go for a second unlock option, the Apple Watch, and unlock your phone. The procedure is identical to that of unlocking a Mac with an Apple Watch. When the unlock occurs, the user receives a haptic buzz and an Apple Watch notification indicating that the unlocking operation was successful. It's worth remembering that your Apple Watch can only be used to unlock your iPhone while wearing a mask - it can't be used to authenticate Apple Pay or App Store transactions.

You must be running iOS 14.5 or later on your iPhone and watchOS 7.4 or later on your Apple Watch for the feature to be accessible. In terms of hardware, you'll need an Apple Watch Series 3 or later, as well as an iPhone X or later with Face ID.

On your iPhone, go to Settings -> Face ID and Passcode and enable the "Unlock iPhone With Apple Watch" option. To use your Apple Watch to unlock your iPhone, make sure it's close and on your wrist, and that it's unlocked using your passcode.

When you first try to unlock your iPhone with Apple Watch while wearing a mask, it will prompt you to input your passcode. Once you've done that, you should be able to unlock your iPhone while wearing a mask (a mask is required - it won't work otherwise). Similarly, if you've removed and replaced your watch, you'll need to re-enter your passcode, or it won't work.

If you meet all of those requirements but still can't obtain Unlock iPhone, the following tips may assist you in getting your Apple Watch to operate.

If all of those requirements are satisfied, but you still can't get Unlock iPhone With Apple Watch to function, the following tips may assist.

Check to see whether your Apple Watch is communicating with your iPhone.

Is your Apple Watch actively linked to your iPhone even though it is paired with it? Swiping up from the bottom of the screen to bring up the Control Center allows you to conveniently check on your Apple Watch. If you see a green iPhone symbol in the top-left corner, your watch has been successfully connected to it.

If the green indicator isn't visible, make sure Bluetooth is turned on in your iPhone (Settings -> Bluetooth) and that your Apple Watch is shown as linked in the "My Devices" list. Disable "Unlock With iPhone" using Apple Watch.

On Apple Watch, watchOS has a setting that allows your iPhone to unlock your watch as long as your iPhone is unlocked (Settings -> Passcode -> Unlock with iPhone).

Some users have discovered that deactivating this functionality and then restarting both devices allow Unlock with Apple Watch to function on iPhone. Of course, because you're turning off one feature to revive another, this is a workaround rather than a remedy, so you'll have to decide whether the compromise is worthwhile in your specific use situation.

Everything about Siri (how to use, where you can use, settings, how to change settings, etc.)

Chapter 9

SIRI

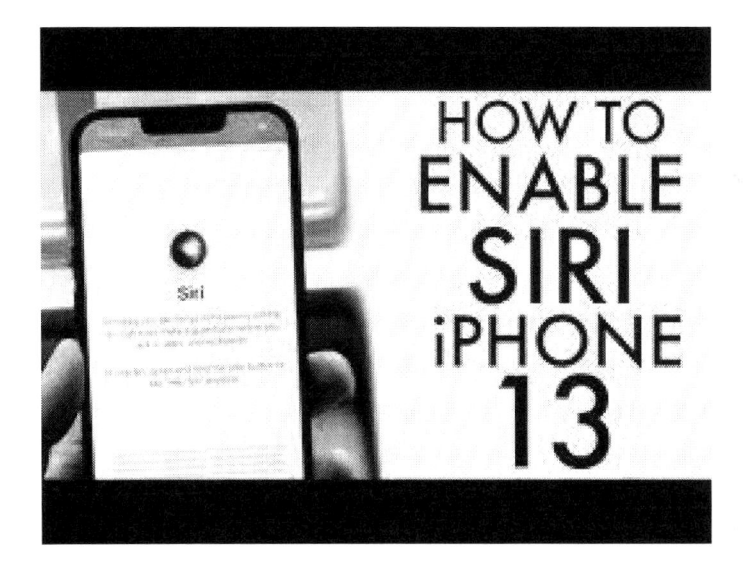

How to Instruct Siri to Control Your HomeKit Devices at a Predetermined Time

For example, if you want your blinds to open at 7 a.m. the next day, you might tell Siri, "Hey Siri, open the blinds at 7 a.m." Siri responds to geolocation instructions as well, so you can say things like, "Hey Siri, turn off the lights when I leave."

When you ask Siri to operate a HomeKit product in this manner, automation is created in the Home app's "Automation" section. If you wish to delete an Automation generated by Siri in the Home app, simply slide left and hit Delete.

In iOS 15, HomeKit developers may also add Siri functionality to their goods. It should be noted, however, that using Siri commands with third-party devices necessitates the ownership of a HomePod to pass the requests through.

Third-party HomeKit gadgets that have Siri integration can be managed using Siri commands for tasks like scheduling reminders, controlling devices, broadcasting messages, and more.

How to Make Siri Read Your Notifications

In iOS 15, here's how you get Siri to announce alerts.

Open the Settings app.

Select Notifications.

Choose Announce Notifications via the "Siri" menu.

Toggle the switch next to Announce Notifications to the ON position in the green.

To have Siri broadcast all notifications from a single app, just choose it from the list under "Announce Notifications From" and enable the Announce Notifications option.

How to Make a Private 'Hide My Email' Address

With iPhone or iPad, choose the Settings app.

At the main settings menu, choose your Apple ID name.

Select iCloud.

Select Hide My Email.

Tap the Create new address button.

Continue, then give your address a unique label. You may also make a note of it if you like.

Then, touch Next, and finally, tap Done.

In Safari, learn how to hide your IP address from trackers.

With iPhone or iPad, choose the Settings app.

Scroll to the bottom and select Safari.

Scroll down to the "Privacy and Security" section and choose 'Hide IP address.'

Trackers and Websites or Trackers Only are the options.

Make the Most of Apple's New Weather Maps

In Apple's native Weather app, three full-screen weather maps are offered. They provide a birds-eye perspective of local precipitation, air quality, and temperature forecasts.

It should be noted that air quality data is only available for Canada, China mainland, Germany, France, India, Italy, Mexico, the Netherlands, South Korea, Spain, the United Kingdom, and the United States.

Tap the little folded map symbol in the bottom-left corner of the Weather app, or scroll down on a forecast page, press on the default temperature map, and then tap on the stack to change the view to precipitation or air quality.

The precipitation maps are dynamic, displaying the direction of oncoming storms as well as the severity of rain and snow. The 12-hour forecast is displayed in the progress bar at the bottom, which you may stop using the Pause button or scrub through by moving the progress dot with your finger.

IPHONE 13 USER GUIDE

The stack icon in the top right corner may be touched to transition to the air quality or temperature maps, which also show you the conditions in your immediate vicinity and the neighboring locations. The icon above it allows you to move between geographical regions in your prediction list, while the top icon zooms in on your selected location.

How to Use the Translate App's Auto-Translation

To activate Auto Translate, first enter conversation mode by tapping the Conversation tab, which can be found at the bottom of the screen in both landscape and portrait mode.

Bottom-right, tap the ellipsis (three dots) icon.

From the popup menu, choose Auto Translate.

The Translate software will now recognize when you start and finish speaking, allowing the other person to react without having to interact with the iPhone.

How to Refresh a Webpage Quickly in Safari

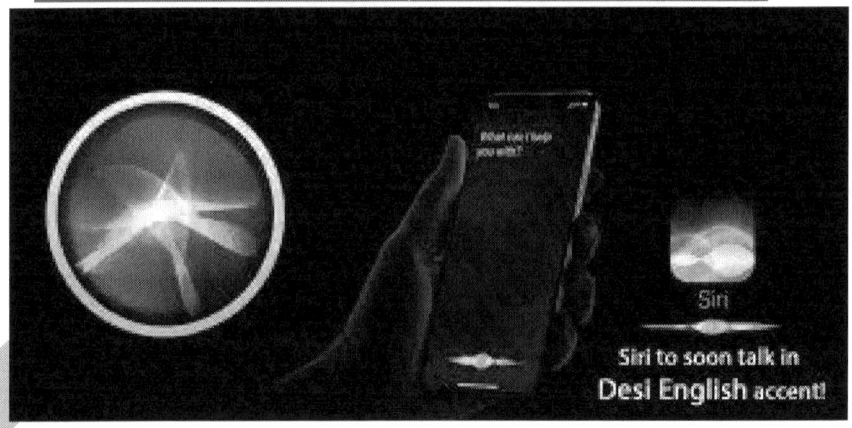

Apple still provides a refresh symbol in the address bar that you can press to reload the currently viewed website.

However, there is now another, less visible option to refresh web pages that you may find more convenient.

In Safari, a downward swipe on any webpage is all that is required to refresh it. This alternative to hitting the reload button is particularly handy if you like to maintain the address bar at the top of the screen, where tapping the reload icon might be inconvenient.

How to Change the Start Page and Background of Safari

With iPhone or iPad, enter Safari.

In the bottom right corner of the Safari screen, tap the open tabs symbol.

To launch a new tab in the Tabs view, press the 'Plus' symbol in the bottom left corner.

Scroll to the bottom of the Start Page and select the Edit button.

Turn on the button next to Use Start Page on All Devices to sync your Start Page settings with additional devices linked to the same Apple ID.

Control what appears on your Start Page by using the switches. In Safari, these are the topics on the front page like Favorites, Frequently Visited, Shared with You, Privacy Report, Siri Suggestions, Reading List, and iCloud Tabs are part of the options one can view.

By pressing the large Plus icon, you can also enable the Background image option and pick one of the available iOS wallpapers or create your own from your photographs. When you're finished, tap the X in the top-right corner of the menu card.

How to Make Use of Tab Groups in Safari

Tap the Open Tabs icon in the bottom right corner of the screen to launch Safari.

Tap or long-press the tab bar at the bottom of the screen in the center.

Choose New Empty Tab Group. Alternatively, if you already have the tabs you wish to combine open, choose New Tab Group from X Tabs.

Enter a name for your Tab Group and then press Save.

After you've made a Tab Group (or several), you may quickly switch between them by pressing the Tab bar in the open tabs view and selecting the one you want. When a Tab Group is selected, all tabs that are opened will be immediately added to that group.

Inquire of Siri

Communicating with Siri is a convenient method to do tasks. Request Siri to translate a word, set an alarm, locate a place, and provide weather information, among other things.

Utilize Your Voice to Activate Siri

Siri responds out loud when you activate it with your voice.

Say "Hey Siri," and then ask Siri a question or assign her a task.

Say "Hey Siri" again or press the Listen button to ask Siri another question or do another action.

Note: To prevent your iPhone from responding to "Hey Siri," face it down.

Activate Siri Through a Button

While a button is used to activate Siri, Siri replies discreetly when the iPhone is in silent mode. Siri responds audibly when quiet mode is disabled. To modify this, see Modify Siri's response.

Keep the side button pressed and held.

When Siri appears, you may ask her a question or have her do a job for you.

Tap the Listen button to ask Siri another question or do another activity.

Correct Siri If She Misunderstands You

Replying to your request: Tap the Listen button, then rephrase your request.

Explanation of a portion of your request: Repeat your request after pressing the Listen button and writing out any unfamiliar words or phrases Siri didn't comprehend the first time. If you want to call someone, say "Call" and then their name.

Before sending a message, modify it as follows: Declare, "Change it."

You can amend your request using text if you see it onscreen. Then, using the onscreen keyboard, tap the request.

Add Shortcuts for Siri

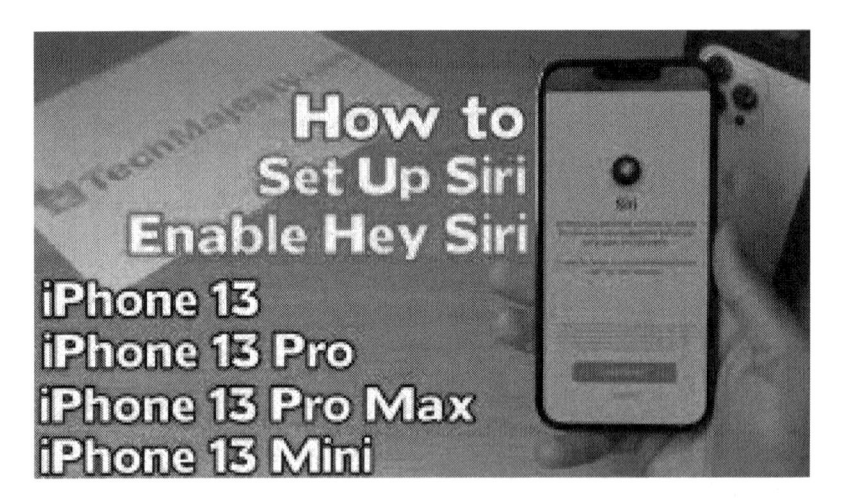

Certain programs include shortcuts for tasks you perform regularly. Siri may be used to invoke these shortcuts using only your voice.

Include a Recommended Shortcut

Shortcut recommendations appear when you tap Add to Siri and apply the on-screen directions to record a word that implements the shortcut when you say it out loud.

Additionally, you may use the Shortcuts app to create a new Siri-enabled shortcut or to manage, re-record, and remove existing Siri-enabled shortcuts.

Utilize a Shortcut

Activate Siri, and then say your shortcut phrase.

How to Use Siri without Having Internet Access

Apple transferred all Siri voice processing and customization to the smartphone in iOS 15, making the virtual assistant more secure and responsive. Additionally, Siri can now fulfill a variety of queries completely offline.

Once you've upgraded to iOS 15, there's no need to activate anything to let Siri operate offline. It is capable of handling the following sorts of requests without communicating with Apple's servers:

Timer and alerts can be created and disabled.

Launch applications.

Control the playing of Apple Music and Podcasts audio.

Change the system's settings, such as the ability to have accessibility features like Audio Routing, Voice Volume, Low Power Mode, and Airplane Mode.

A response like "You need to be online to do that" or "I can help with that once you're connected to the internet" will appear if you ask Siri to do something that requires internet access but you don't have data or Wi-Fi connection. Examples include sending a message, checking the weather, or playing streamed content.

Charge iPhone With MagSafe Charger

Connect the MagSafe Charger to the power supply through the Apple USB-C 20W power adapter or any compatible power adapter.

The charger should be attached to the iPhone or its MagSafe case or sleeve via USB cable. As soon as your iPhone starts charging, you'll see a charging icon on your screen.

Connect Your Apple Watch to Your iPhone

Tap the Apple Watch app on your iPhone, then follow the on-screen instructions, and follow the instructions.

Unlock Your iPhone with Your Apple Watch

When worn with a face mask, you may use your Apple Watch to securely unlock your iPhone.

To enable Apple Watch to unlock your iPhone, take these steps:

Navigate to Settings > Face ID and Passcode.

Scroll down and then activate Apple Watch (below Unlock with Apple Watch).

Turn on the setting for each watch if you have more than one.

While wearing an Apple Watch and a face mask, raise or push the iPhone screen to wake it, then gaze at it to unlock it.

Unlocking your iPhone requires that you have a passcode on your Apple Watch, that it is unlocked and, on your wrist, and that it is nearby your iPhone.

Chapter 10

HANDOFF WORK ACROSS YOUR IPHONE AND OTHER DEVICES

Through the use of Handoff, you are able to start anything on one device (such as your iPhone, iPad, iPod touch, or Mac) and finish it on another (iPhone, iPad, iPod touch, Mac, or Apple Watch).

For instance, you may begin an email response on your iPhone and finish it in Mail on your Mac. Handoff is compatible with a wide variety of Apple applications, including Calendar, Contacts, and Safari. Certain third-party applications may also be compatible with Handoff.

Prior to Beginning

To ensure that tasks are transferred between your iPhone and another device, ensure the following:

Both devices have the same Apple ID associated with them.

Your gadgets are linked to the Internet through Wi-Fi.

Your Bluetooth-enabled gadgets are within range of one another (about 33 feet or 10 meters).

Handoff is enabled in System Preferences > General, and Bluetooth is enabled in System Preferences > Bluetooth on your Mac.

Handoff is enabled under Settings > General > AirPlay and Handoff on your iPhone and another iOS or iPadOS device, and Bluetooth is enabled in Settings.

Each device is running the latest version of the necessary software: iOS 10, iPadOS 13, macOS 10.10, or watchOS 1.0.

Handoff from a Different Device to Your iPhone

On your iPhone, open the App Switcher. At the bottom of the iPhone screen, the Handoff symbol for the app you're now using on your other device displays.

To continue working in the app, tap the Handoff symbol.

Handoff From The iPhone to Another Device

To continue working in the app on the other device, click or press the Handoff icon.

On other devices, the Handoff symbol for the app you're now using on iPhone appears in the following locations:

Mac: The Dock's right-hand side (or at the bottom, depending on the Dock position).

iPad: The dock's right end.

At the bottom of the App Switcher screen on an iPhone or iPod touch.

Disable Handoff on All of Your Devices

iPhone, iPad, and iPod touch: Navigate to Settings > General > AirPlay and Handoff.

Turn off "Allow Handoff between this Mac and your iCloud devices" on the Mac by selecting Apple Menu > System Preferences > General.

Between Your iPhone and Other Devices, Cut, Copy, and Paste

A block of text or an image, for example, may be copied and pasted from your iPhone to an iPad or another iOS device as well as to a Mac using Universal Clipboard.

Prior to the Start

When cutting or copying and pasting between your iPhone and another device, take the following precautions:

Both devices have the same Apple ID associated with them.

Wi-Fi is enabled on your devices.

Bluetooth connectivity exists between your gadgets (about 33 feet or 10 meters).

Handoff is enabled under Settings > General > AirPlay and Handoff on the iPhone and another iOS or iPadOS device, and Bluetooth is enabled in Settings.

On a Mac, turn on Handoff via System Preferences > General and Bluetooth via System Preferences > Bluetooth.

Each device is running the latest version of iOS, iPadOS, or macOS: iOS 10, iPadOS 13, or macOS 10.12.

Copies, Cuts, and Pastes

Pinch three fingers together to form a seal.

Pinch three fingers together twice to seal the cut.

Using three fingers, pinch open the paste.

Additionally, you may pick an item by touching and holding it, then tapping Cut, Copy, or Paste.

Important: You only have a limited amount of time to cut, copy, and paste your text.

Sync Your iPhone to Your PC

You can use iCloud to automatically sync all of your Apple ID-enabled devices' photos, files, calendars, and other data. (You can even use a Windows PC to access your iCloud data on iCloud.com.) You can access more content across all of your devices with third-party services like Apple Music.

The following objects can be synced with your iPhone if you don't want to use iCloud or any other services.

Albums, singles, playlists, films, television programs, podcasts, audiobooks, novels, photographs, and video clips.

Calendars and contacts

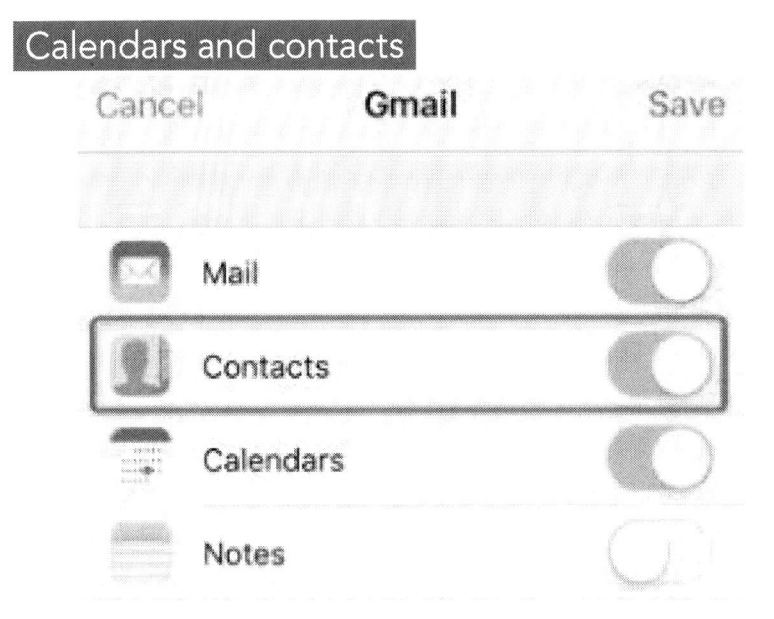

[Screenshot showing Gmail account settings with toggles for Mail, Contacts, Calendars, and Notes — Contacts highlighted]

You may sync these objects between your computer and iPhone to keep them current.

If you use iCloud or other cloud-based services like Apple Music, the option to sync with your computer may be unavailable.

Configure Your Mac and iPhone For Synchronization

A cable is required to connect your iPhone to your computer.

Select your iPhone from the Finder sidebar on your Mac.

Select the type of content you wish to sync at the top of the window (for example, Movies or Books).

Select Sync [content type] to [device name] from the drop-down menu.

You may choose to just sync some items, such as music, movies, books, or calendars, instead of the entire content category.

Rep the previous steps 3 and 4 for each type of material you wish to sync, then click Apply.

How to Send Content from an iPhone to a Computer

iCloud Drive allows you to keep your files up to current and available across all of your devices, including Windows PCs. You can also use AirDrop to transfer files between iPhones and other devices, as well as send email attachments.

How to Send Files with iPhone to Mac

Connect your iPhone to your computer.

You can connect by USB or, if Wi-Fi syncing is enabled, over a Wi-Fi connection.

Select your iPhone from the Finder sidebar on your Mac.

Click Files at the top of the Finder window, then one of the following:

Move a file or group of folders from a Finder window onto the name of an application in the queue to transfer between Mac to iPhone.

Transferring files from iPhone to Mac: Move a file from your iPhone's files list to a Finder window by clicking the disclosure triangle next to the application's name.

How to Move File with iPhone to PC

Install or upgrade iTunes on your PC to the most recent version.

Connect your iPhone to a Windows PC with the included USB cable.

IPHONE 13 USER GUIDE

You can connect by USB or, if Wi-Fi syncing is enabled, over a Wi-Fi connection.

Click the iPhone button in the top left of the iTunes window on your Windows PC.

Click File Sharing, then choose an app from the list and do one of the following:

The following steps describe how to move a file from iPhone to your computer: From the list on the right, choose the file you wish to transfer, next, click "Store to," go to the location where you wish to save the file, and afterward click Save To.

To transfer a file from your computer to your iPhone, follow these steps: Click Add, then choose the file you wish to transfer.

> Connect your iPhone and a computer using the USB cable for a variety of reasons.

- iPhone set up
- Charge the iPhone
- Internet connection Hotspot
- File transfer

Plugin the USB cable to the iPhone and the computer's USB port either directly or through a USB adapter.

Handoff

The Handoff feature enables you to continue on another device where you stopped on another device. Handing off tasks is compatible with a lot of Apple apps like Mail, Safari, Pages, Numbers, Keynote, Maps, Messages, Reminders, Calendar, among others.

IPHONE 13 USER GUIDE

The devices should have Bluetooth turned on and be in the range of about 10 meters.

Continuity with Handoff: Once you are logged into the MacBook Pro, iOS device or iPadOS devices with the same Apple ID and WiFi and Bluetooth is enabled, an icon will appear in the Dock whenever an activity is being handed off. You can be checking out information on a webpage on your MacBook Pro and Handoff to your iPhone to pick up where you have left off. Handoff works with a lot of Apple Apps, but the devices involved must meet certain system requirements. To turn Handoff on, go to System Preferences > General > Allow Handoff this Mac and your iCloud devices.

Turn off Handoff: To turn 'Handoff' off on your Mac, go to System Preferences > General > Allow Handoff this Mac and your iCloud devices but select deselect the option 'Allow Handoff this Mac and your iCloud devices.' On your iPhone, iPad, and iPad Touch, Settings > General > AirPlay and Handoff to disable Handoff.

Manage Airplane mode: When traveling with the iPhone or in a place where wireless communication is prohibited, you can keep the device turned on but in airplane mode. While in airplane mode, you can take notes, listen to music, watch movies, read books, play games, and do other things that do not require internet access. Go to Control center > Airplane mode (and click to turn on Bluetooth or Wi-Fi). You can turn it off when you want to exit the airplane mode.

Erase iPhone

To remove all your data, content, and settings from your iPhone even after deleting them, you may need to erase the iPhone. Go to Settings > General > Reset > Enter Passcode > Erase All Content and Settings.

Before passing your iPhone to someone else to use, it is important to back up and erase the iPhone

Restore to iPhone to Default Settings

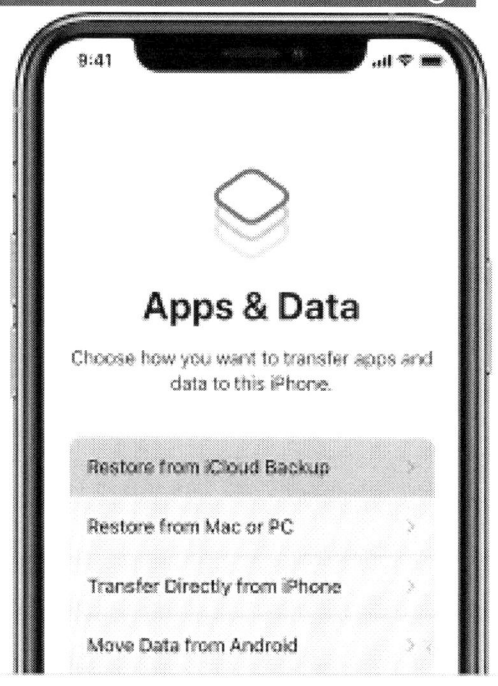

You can restore your iPhone to the factory settings without necessarily erasing the content. Go to Settings > General > Reset > Reset All Settings. This would reset network settings, dictionary of the keyboard, location settings, privacy settings, and Apply Pay settings. The data and media stored on the phone are not erased or deleted.

You can also reset specific settings and not the entire settings on the iPhone. Settings such as network settings, keyboard dictionary, home screen layout, and location and privacy.

IPHONE 13 USER GUIDE

Back up your iPhone

Backing up your iPhone helps to ensure that you don't lose important files, pictures, documents, and content. You can either back up to your computer or iCloud.

To back up using iCloud, navigate to Settings > {Your name} > iCloud > iCloud Backup > Turn on iCloud Backup. iCloud will back up by default when the device is locked and connected to power and Wi-Fi. However, if you would like to complete a manual backup at any time, Settings > {Your name} > iCloud > iCloud Backup > Back Up Now.

To back up using a Mac, ensure that the iPhone is connected to it via USB. Go to the Finder on the Mac and click on your iPhone icon. Then, General > Back up all of the data on your iPhone to this Mac > Encrypt local backup > Back Up Now.

Apps and Features of the iPhone

Many great and wonderful apps are preloaded on the iPhone that will boost your productivity and help you do a wide variety of things conveniently, easier and faster. The following apps and features come right out of the box with the iPhone or are available in the AppStore.

Airdrop

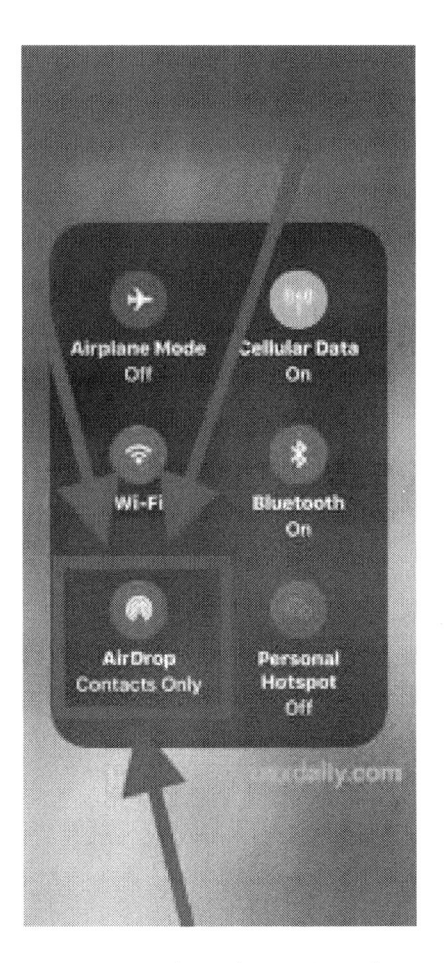

Airdrop enables you to send and receive documents, photos, map locations, files, and webpages wirelessly (Wi-Fi and Bluetooth) to a nearby Mac, iPhone, or iPad.

Share files via Airdrop: Just click on the item you want to send, select the Share > Airdrop, and then the profile picture of a nearby Airdrop user. From the Finder app, click Airdrop and drag the file to the recipients' device. The recipient has to choose whether or not to accept the file for the share to be completed.

Receive files via Airdrop: Ensure that in the AirDrop window, you have set 'Allow me to be discovered by' and choose the appropriate option (no one, contacts, or everyone) in Control Center. Navigate to the Airdrop notification and click Accept from the pop-up menu. The file received will be added to the Downloads folder by default.

Airdrop works over Bluetooth, and as such, the Bluetooth of both devices must be turned on and within 30 feet (9 meters) of each other.

Chapter 11

HOW TO USE HIDE MY EMAIL TO CREATE AN EMAIL ADDRESS

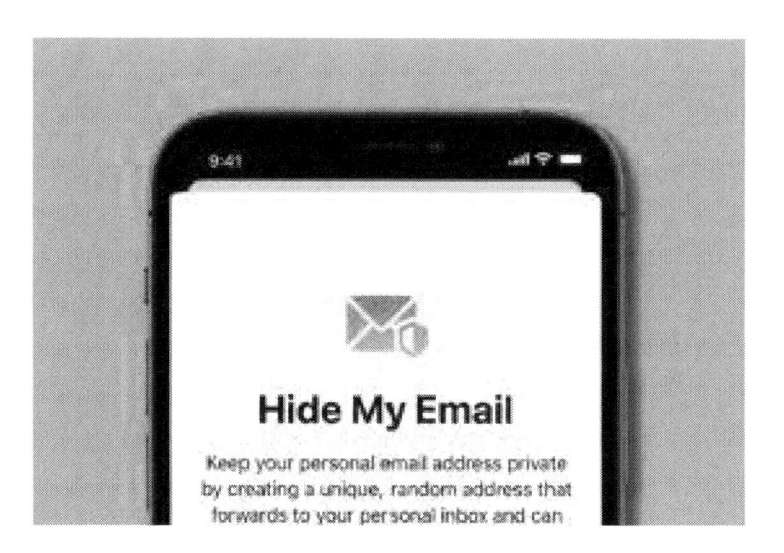

The following instructions demonstrate how to utilize Hide My Email to establish a new fake email address for usage in Safari and Mail. Ascertain that your iOS device is running iOS 15 or a later version.

Go over to the Settings application.

At the top of the main settings menu, tap your Apple ID name.

IPHONE 13 USER GUIDE

Select iCloud.

Select Hide My Email from the drop-down menu.

Create a new address by tapping.

Continue, then provide an identifying label for your address. Additionally, you may add a remark about it.

Next, then Done.

You may now send emails using the random email address in Mail or when prompted to input your email address on a website in Safari.

How to Keep Your IP Address Private While Using Safari

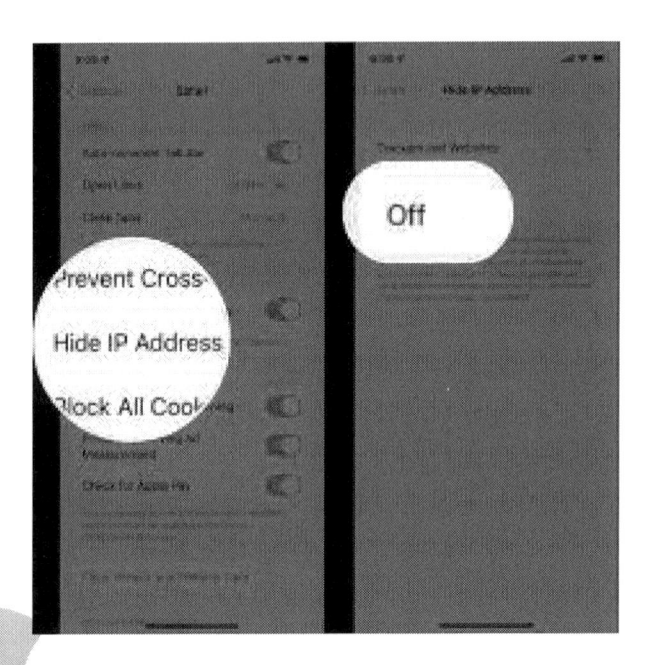

In iOS 15, Apple has enhanced Safari's Intelligent Tracking Prevention feature to protect you from web trackers.

Apple began introducing Intelligent Tracking Prevention in 2017, a privacy-focused feature that makes it more difficult for websites to follow users throughout the web, prohibiting the creation of browser profiles and histories.

Rather than blocking ads, Intelligent Tracking Prevention instead prohibits websites from knowing who you are and following your online browsing history without your permission. Apple has furthermore granted you the ability to mask your IP address from them. Activating the function is just a matter of following these steps.

Go over to the Settings application.

Scroll to the bottom and hit Safari.

Scroll down and hit Hide IP address under the "Privacy and Security" section.

Choose between Trackers and Websites or just Trackers.

Safari will block trackers and websites from learning your IP address, should you have set up the new iCloud Private Relay feature in iOS 15. But if you don't have a paid iCloud Plus account, it is possible to hide your IP address from trackers even using the other option.

IPHONE 13 USER GUIDE

How to Activate and Deactivate iCloud's Private Relay

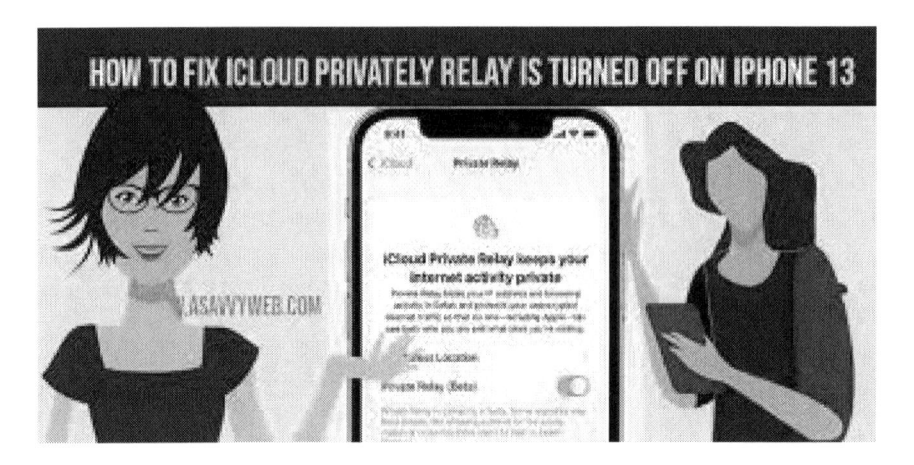

Along with iOS 15, Apple announced the iCloud+ service, which enhances the capabilities of Apple's premium iCloud plans (upgraded iCloud storage tiers begin at $0.99). One of these capabilities is iCloud Private Relay, which is intended to encrypt any communication leaving your device in order to prevent it from being intercepted or read.

To route your web traffic through a private server that removes your IP address, connect through an Apple-managed server. When the IP address is removed, Apple routes the traffic to a second server, which has been maintained by a third-party company, to a different IP address, which is used temporarily before being returned to its intended destination. Thus, Apple blocks any potential data that could be used to profile your identity, along with your IP address, location, and online activities.

Involving a third party in the relay system is a deliberate decision, Apple claims, meant to prevent anyone, even

Apple, from knowing a user identity or the website they are accessing.

The following instructions will walk you through the process of enabling and disabling Private Relay on an iPhone or iPad running iOS 15.

Go over to the Settings application.

At the top of the main settings menu, tap your name.

Select iCloud.

Private Relay should be tapped.

Toggle the switch adjacent to iCloud Private Relay on/off. If you're turning it off, confirm by tapping Turn Off Private Relay.

With Private Relay enabled, you may toggle between the default Maintain General Location option and the less geographically detailed and more private Use Country and Time Zone option by selecting IP Address Location.

IPHONE 13 USER GUIDE

CONCLUSION

In 2020, Apple revamped many aspects of its iPhone design. They conveyed back the square-shaped plan of the iPhone 4 model, altered the back camera module plan, and made different refinements. He likewise presented the iPhone 12 Mini. This year, Apple left things pretty much the same. Of course, the phone fronts are a bit smaller, the internals are better, and there are lots of new features. Generally, however, on the off chance that you currently own an iPhone 12 device, there's likely insufficient here to warrant a 1-year update.

Actual measurements to the side, the iPhone 13 and the Mini model are essentially similar telephone devices.

The iPhone 13 is also equipped with new technology that Apple calls ProMotion. In the iPhone 13 camera, a 12 MP pixel sensor is used instead of the older 16 MP pixel sensor. So, it is actually the same sensor, just that it has a different type of display. What this means is that the pixel size of the sensor is the same as the pixel size on the front camera. And it is 1.4 μm, which is the same pixel size as the Google Pixel 3. This means the iPhone 13 camera is capable of very detailed, high-resolution images.

The iPhone 13 series must be among the company's best devices for photography and videography. However, the updates here aren't as drastic as someone with an iPhone 12, or even an iPhone 11 would need to do.

The Apple iPhone 13 series is powered by Apple's latest and

greatest silicon, known as A15 Bionic. It is a 5 nm+ construct. Although we haven't had a chance to compare this processor yet, we are quite confident that it will be the fastest mobile processor on the market at the moment.

As such, performance won't be an issue with iPhone 13 models. They should be able to handle just about any task you assign to them. Pro models, however, have more RAM than non-Pro models, so they'll be a bit better for heavy-duty tasks, especially if you have a lot of background apps.

The A15 Bionic is expected to be a bit more energy efficient than last year's A14 Bionic. This means that we can see a better battery life of the iPhone 13 models even when you factor in the bigger batteries.

All new iPhones come with iOS 15 on board. This is the most recent rendition of the working framework. Overall, iPhones have seen around 5 years of software updates. This means that you should expect an annual update to the latest version of iOS by 2026 on any of the iPhone 13 models.

Made in the USA
Middletown, DE
15 February 2022